T0353205

How Things Work

How Things Work

Series Editor: Charles F. Bowman

How Things Work: The Computer Science Edition

Charles F. Bowman

For more information about this series please visit: https://www.routledge.com/How-Things-Work/book-series/HTW

How Things Work

The Computer Science Edition

Charles F. Bowman

CRC Press
Taylor & Francis Group
Boca Raton London New York

CRC Press is an imprint of the
Taylor & Francis Group, an **informa** business

A CHAPMAN & HALL BOOK

First edition published 2022
by CRC Press
6000 Broken Sound Parkway NW, Suite 300, Boca Raton, FL 33487-2742

and by CRC Press
2 Park Square, Milton Park, Abingdon, Oxon, OX14 4RN

© 2022 Plenum Press, LLC

CRC Press is an imprint of Taylor & Francis Group, LLC

Library of Congress Cataloging-in-Publication Data

Names: Bowman, Charles F., author.
Title: How things work : the computer science edition / Charles F. Bowman.
Description: First edition. | Boca Raton, FL : CRC Press, 2022. | Series:
  How things work | Includes bibliographical references and index.
Identifiers: LCCN 2021012185 | ISBN 9780367568375 (hbk) | ISBN
  9780367567316 (pbk) | ISBN 9781003099567 (ebk)
Subjects: LCSH: Computer science--Popular works.
Classification: LCC QA76 .B686 2022 | DDC 004--dc23
LC record available at https://lccn.loc.gov/2021012185

ISBN: 978-0-367-56837-5 (hbk)
ISBN: 978-0-367-56731-6 (pbk)
ISBN: 978-1-003-09956-7 (ebk)

Typeset in Minion
by KnowledgeWorks Global Ltd.

Access the Support Material: https://www.routledge.com/9780367568375

Other Books by Charles F. Bowman

Algorithms & Data Structures: An Approach in C
Objectifying Motif
Broadway: The New X Windows System
The Wisdom of the Gurus

For Florence,
Your Melody Remains

For Maria,
The Sunrise of My Second Dawn

For Nicole, Michael, and Charles,
If Pride is Truly a Sin, I will Surely Burn in Hell

For Amaya, Joseph, Nora, CJ, and Mathew,
I Hope the Footprints I Leave, Lead You on the Right Path

Contents

Acknowledgments

Although only one name appears on the cover, a project of this size requires the contributions of many dedicated professionals. I am grateful to all of them.

I'd like to begin by thanking my editor, Randi Cohen, for her patience, suggestions, and making the process of authoring a book as pleasant as possible.

I would also like to extend a special thank you to Talitha Duncan-Todd. Her efforts ensured that the speedbumps were few and far between.

During development, a manuscript undergoes many critiques. The suggestions of the following reviewers made this book a better read:

- My friends and business partners Michael Bardash and Joseph Cerasani

- Members of the Mahwah Writers Group

- The CRC reviewers

That said, any errors or omissions are mine alone.

The wonderful staff at CRC Press helped shepherd the manuscript through all the phases of production. Their dedicated professionalism is evident in the quality of the book. It was a pleasure to work with them.

- Ashraf Reza

- Ed Curtis

Finally, I'd like to thank my children Charles, Michael, and Nicole for their support, and Maria for always being there when I need her.

Author

Charles F. Bowman is a respected Senior Software Architect serving numerous prestigious clients. He has taught both graduate and undergraduate computer science courses at St. John's University, the City University of New York, and St. Thomas Aquinas College.

Mr. Bowman has published several books, including *Algorithms and Data Structures: An Approach in C* (Harcourt Brace/Oxford University Press); *Objectifying Motif* (Cambridge University Press/SIGs Books); *Wisdom of the Gurus* (Cambridge University Press/SIGs Books); and *Broadway: The Complete Internet Architecture* (Addison/Wesley).

Mr. Bowman has served as Editor-in-Chief for *The X Journal, UNIX Developer,* and *CORBA Development* and Series Editor for the Managing Object Technology book series of Cambridge University Press/SIGs Books. He is a regular contributor to many respected journals and magazines.

A graduate of New York's prestigious Brooklyn Technical High School, Mr. Bowman earned a BS in computer science at St. John's University and an MS in computer science at New York University.

CHAPTER 1

Computers Are Everywhere

I do not fear computers. I fear the lack of them.

ISAAC ASIMOV

INTRODUCTION

Depending on your perspective, computers are either the bane or the blessing of modern life. They help us file our taxes, guide doctors through sophisticated surgeries, and deliver entertainment on demand. Unfortunately, they can also overwhelm us with techno-jargon, drown us in a tsunami of spam and unwanted advertising, and serve as targets for identity theft. Nonetheless, can you envision life without them?

Like any other tool, the more you learn about computers, the more useful they become. However, most of us have neither the time nor the inclination to become computer programmers or software architects. Nonetheless, we certainly don't want to remain wholly uninformed so that every mouse click seems like "magic." Thus, that begs the question: How much technology does someone need to understand to harness the power of the electronic world?

That is the very intent of this book: to arm you, the reader, with enough information to remove the mystique and apprehension associated with computers and replace it with a broad understanding of their inner workings so that your interaction with them becomes more rewarding and less frustrating.

MODERN TECHNOLOGY

I'm sure everyone reading this book owns or is familiar with many of the following products:

- Smartphone

- Smartwatch

- Smart TV

- eBooks/eReaders (e.g., Kindle)

- Smart Thermostats

- Navigation Systems (GPS)

- Smart Assistants (e.g., Alexa, Google Home)

Smart devices like those listed above have all found hooks on the toolbelts of modern life. Consider: When was the last time you asked for driving directions or "looked up a word" in a printed dictionary? Do you ask Alexa or Google Home for the weather report or a daily news briefing? Does your refrigerator automatically order milk from your online grocer when you're running low?

As the previous examples demonstrate, interaction with computers is unavoidable in today's world. Sometimes it's obvious. For instance, it's clear that you're using an "intelligent" device when you power up your laptop. But how about when you press a button on an elevator? Or when you change gears in your car?

Whether we're comfortable with computers or not, we need to accept them—even embrace them. Still, many folks opt to remain ignorant of what is a fundamental aspect of modern living. Such an approach is not consistent with other facets of our lives, however.

For example, most car owners don't know how to design automobiles. Yet, even if they don't understand the details, they are familiar with an automobile's major components: engine, transmission, suspension, etc. Though minimal, such familiarity with automotive design makes drivers more comfortable sitting behind the wheel. Nonetheless, despite our general acceptance of technology, many of us turn our heads and wave our hands when computers are the topic of discussion.

Unfortunately, in the world of technology, ignorance is not bliss. What you don't know might wind up hurting you.

OBJECTIVES OF THIS BOOK

I'm sure many readers of this book have no desire to become electrical engineers or software developers—you'd be reading different types of books. That's good because this text aims to help layfolk and students understand—at a conceptual level—the inner workings of computer systems. As we progress, we'll learn what this incredible technology can do *for* us. Additionally, in Chapter 13, we'll also discover what it can do *to* us.

Throughout the text, we'll use plain everyday language. Nonetheless, to become conversant in the world of computing, you'll need to understand some basic terminology. However, whenever jargon is unavoidable, we'll introduce it delicately and explain it thoroughly.

ORGANIZATION

From a pedagogical perspective, the book will introduce each main subject (hardware, software, networking, etc.) with an introductory chapter that presents the material in broad strokes. Subsequent chapters will dive deeper into the subject matter.

Our journey begins with a brief history of the evolution of computers, followed by an overview of *digitization*—one of the most fundamental facets of this technology. After that, we'll describe the hardware components used in modern-day computers and discuss how systems "talk to each other" (i.e., *networking*).

Given that foundation, we'll address the most challenging subject in computer science: software. After reading those chapters, you'll have a solid understanding of what software is (and what it's not!).

Once we understand how hardware and software work together, we'll describe how computer programs execute and how software developers build them. Then, we'll put all this knowledge to work and write some code. After completing this section, you'll feel much more comfortable working with new applications in the future.

Unfortunately, no discussion of the modern computing world would be complete without reviewing its darker side. Thus, this book will inform you about the dangers that lurk behind the lure of the Internet, the risks behind the façade of free software, and the precautions you can adopt to protect yourself.

All technical chapters will include an *Advanced Topic* section, which presents the chapter's main ideas in ways that are, well, more advanced. And there's also a glossary of terms containing definitions for all the terms and acronyms used throughout the text.

So, hop on the train—but leave the baggage of your apprehensions behind—because we are about to embark on an enjoyable journey of discovery.

FORMATTING CONVENTIONS

To simplify the presentation of this material, we've adopted the following formatting conventions:

- When introduced, all technical jargon will appear in *italics*.

- All computer-related terms—file names, command-line text, and program output—will appear in THIS FONT.

- We'll highlight formal definitions using centered text highlighted in gray.

> This is an example of a formal definition.

- All program listings are highlighted in gray contain numbered lines, and the code will appear in THIS FONT.

```
001 // THIS IS A SAMPLE CODE LISTING
002 #INCLUDE <STDIO.H>
003 MAIN()
004 {
005      PRINTF( "HELLO WORLD\n");
006 }
```

CHAPTER 2

A Brief History of Computing

Computers are useless. They can only give you answers.

PABLO PICASSO

OVERVIEW

It's often the case that to know where you're going, you need to understand where you've been. The study of technology is no different. Thus, this chapter will demonstrate how, through necessity and ingenuity, these marvelous machines we call computers evolved in a rather pedestrian manner.

For those readers who are not interested in the origins of computing, feel free to skip this chapter.

MAJOR ADVANCES IN COMPUTING

Most human progress is a series of small advances building upon prior accomplishments (and failures). Thus, when examining the historical development of any discipline, it's often difficult to establish its "beginning." Indeed, one can trace the roots of many fields of study to the dawn of civilization. The evolution of computing is no exception in this regard because one could argue that its origins date back to the time humans began "counting."

However, as this is not a formal textbook or doctoral dissertation, we will take some liberties in our review of the history of computers. What follows is a "Cook's tour," highlighting the more notable achievements while omitting much of the details.

Abacus—2700–2300 BCE

We'll begin our discussions with the advent of the *abacus*. Although many of the details of its origin have become obscured in the vagueness of history, chroniclers believe that the abacus emerged sometime during the period 2700–2300 BCE.[1]

---

[1] BCE is an abbreviation for Before [the] Common Era. It's the contemporary equivalent of BC.

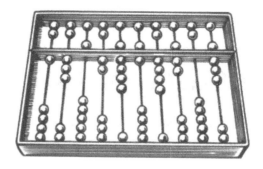

FIGURE 2.1   Abacus.

As depicted in Figure 2.1, an abacus contains arrays of rods and beads representing successive orders of magnitude.[2] Users manipulate the beads to perform basic mathematical calculations such as addition, subtraction, multiplication, and division. Despite its apparent simplicity and the emergence of electronic calculators, the abacus is still in use today.

Babbage's Analytical Engine—1834–1871

In the 1830s, an English mathematician named Charles Babbage developed a mechanical device he called an Analytical Engine (AE) (see Figure 2.2). The AE accepted input (programs and data) using punch cards—a technology commonly used at the time to direct the

FIGURE 2.2   Babbage's Analytical Engine.

---

[2] Like powers of 10.

operations of mechanical looms. The AE would generate output using an early version of a printer[3] and by creating punch cards.

Although never fully implemented, historians regard Babbage's AE as the first computer because it shares many design features found in modern electronic devices: expandable memory, a separate arithmetic component, and logic processing that would have enabled it to execute a programming language that looks remarkably like a modern-day *assembler language*.[4]

The Advent of Software—1840s

During the latter part of 1842 and the early months of 1843, Ada Lovelace (Augusta Ada Byron), an English mathematician and writer, translated an article written by the Italian mathematician Luigi Menabrea that discussed Babbage's Analytical Machine. As part of her translation, Ms. Lovelace also included a method to calculate a sequence of Bernoulli Numbers using the AE[5] (see Figure 2.3). As a result, many historians consider Ada Lovelace the world's first computer programmer and her calculation method as its first computer program.

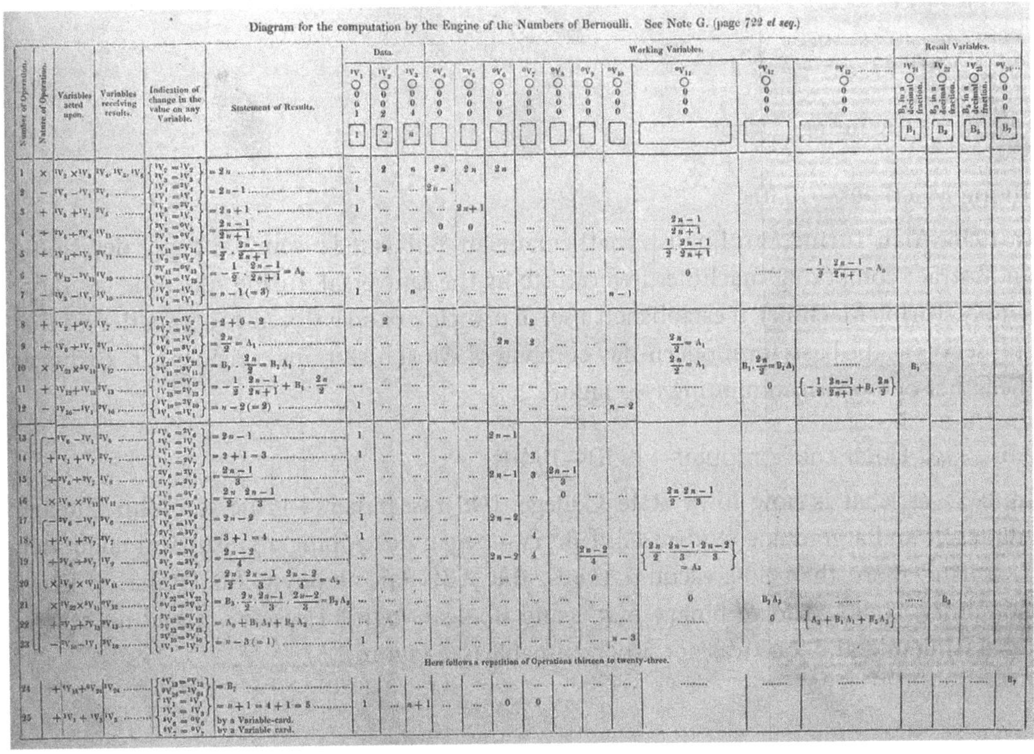

FIGURE 2.3   Diagram produced by Ada Lovelace.

---

[3] Babbage produced one printer during his lifetime.

[4] We will define what an *assembler language* is in Chapter 5.

[5] Bernoulli Numbers are used in number theory.

FIGURE 2.4   A Turing Machine.

Turing Machines—1930s

In 1936, Alan Turing,[6] an English mathematician, published a seminal paper describing an abstract computing machine. Referred to as the Universal Turing Machine (or more simply, Turing Machine), it established a set of principles and a stored-program design that still serves as the basis for modern-day computers. Most historians consider Mr. Turing as the father of modern computing (see Figure 2.4).

ABC: First Electrical Computer—1930s–1940s

In 1937, at what is now Iowa State College, two researchers—Professor John Vincent Atanasoff and a graduate student Cliff Berry—began developing an electrical computer. Requiring more than 300 vacuum tubes, the ABC (an acronym for Atanasoff-Berry Computer) could perform binary math[7] and Boolean logic.[8] However, it had no central processing unit and thus was not programmable (see Figure 2.5).

---

[6] Among his many achievements, Mr. Turing was instrumental in breaking the German Enigma code during WWII. Historians estimate that his work shortened the war by at least two years and saved more than 14 million lives. Nonetheless, despite his acclaim as a war hero, he was prosecuted in 1952 for engaging in "homosexual acts." Sadly, in 1954, less than a month before his 42nd birthday, Mr. Turing died of an apparent suicide.

[7] As we will see in Chapter 3, binary math are calculations performed using binary numbers (0s and 1s).

[8] In the mid-19th century, English mathematician George Boole developed the mathematics of logic, which serves as the foundation for modern electronic circuits.

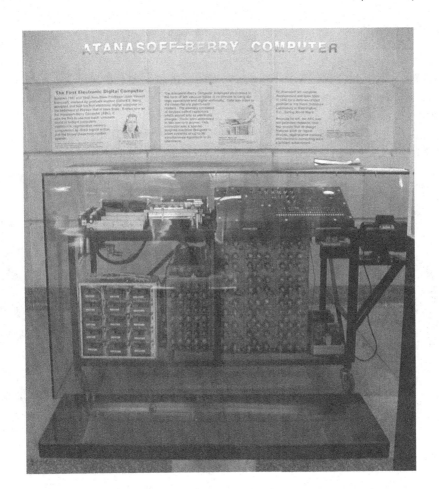

FIGURE 2.5   Atanasoff-Berry Computer.

ENIAC: First Digital Computer—1940s

During the early 1940s, the United States Army funded the construction of the ENIAC (Electronic Numerical Integrator and Computer, pronounced "EEE-NEE-ACK"), the first digital[9] computer. Designed by J. Presper Eckert and John Mauchly at the University of Pennsylvania, the ENIAC became operational in 1945. It contained 20,000 vacuum tubes, weighed over 50 tons, and required about 1,800 square feet of space (see Figure 2.6).

The Advent of the Transistor—1940s–1950s

In 1947, the computer revolution truly began when Bell Laboratories researchers John Bardeen, Walter Brattain, and William Shockley developed the first working *transistors*. You can envision a transistor as a device that acts as a type of electronic switch. Based on its current state, it either permits or inhibits the flow of electrons through it. Engineers can combine transistors to construct the fundamental components that form the foundation of digital electronics.

---

[9] We will discuss what *digital* means in Chapter 3.

FIGURE 2.6   The ENIAC Computer.

Expanding on that discovery, in 1959, Mohamed Atalla and Dawon Kahng invented another type of transistor called a Metal-Oxide-Semiconductor Field-Effect Transistor (MOSFET). The basis of all modern electronics, the MOSFET is the most widely manufactured device in human history (see Figure 2.7).

The First Commercial Computer—1951

After leaving the University of Pennsylvania in 1946, Presper Eckert and John Mauchly (the ENIAC inventors, see above) formed their own company and developed the UNIVersal Automatic Computer I (UNIVAC I), which became the first general-purpose electronic digital computer designed for business. Their first customer was the United States Census Bureau, who took ownership of their UNIVAC on March 31, 1951 (see Figure 2.8).

The First Home Computers—1977

The world changed in 1977—although most people didn't realize it at the time. In that year, Apple, Radio Shack, and Commodore all introduced mass-market computers. Though considered "clunky" by today's standards, these machines brought computing power to the masses. For the first time, home users could leverage the power of technology for a

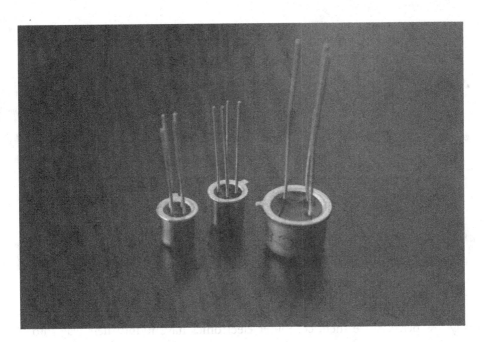

FIGURE 2.7   Replica of the first transistor.

FIGURE 2.8   The UNIVAC I.

FIGURE 2.9 The Big 3 of 1977.

relatively low price. Since then, computer electronics have found their way into almost every facet of human endeavor, revolutionized nearly every science, and fundamentally altered life on this planet (see Figure 2.9).

SUMMARY

The advent of computer technology was more of an evolution than a revolution. There was no magic moment of enlightenment where the heavens parted, trumpets sounded, and spotlights focused on a magical new device.

On the contrary, as with most human endeavors, it required the painstaking rinse-and-repeat process of thought, experimentation, and failure for computers as we know them to come of age.

In Chapter 3, we will start to peek under the hood and understand how electronic devices function.

CHAPTER 3

Digitization

People use technology only to mean digital technology. Technology is actually everything we make.

MARGARET ATWOOD

OVERVIEW

Now that we appreciate how computers evolved, we can begin our understanding of how they work. But first, we need to learn how they consume and process data.

Because computers don't manage information the way humans do, we need to transform data into formats suitable for electronic devices. We call that process *digitization*.

WHAT IS DIGITIZATION?

Though what they accomplish is often remarkable, electronic devices do not possess our senses or cognitive abilities. For example, a computer cannot do what you're doing right now: reading and understanding the text on this page.[1] But consider that the words you're reading appear in an order that you've never seen before.[2] Yet, as you read them, you are interpreting and understanding their meaning.

Extending the above analogy, computers can't admire a painting and infer the artist's message or listen to a symphony and feel the composer's passion. Computers can only perform the tasks that their programming directs them to do. In other words, if human programmers didn't include the instructions to execute a given feature, the computer can't perform it.

Because computers don't possess our senses, we need to transform human-consumable data into a format suitable for electronic processing. For reasons that we'll understand later in the text, the language of computers is *digital*: only 1s and 0s. Whether it be the Oscar-winning movie you streamed the other night, the vacation photographs you downloaded

---

[1] This statement is not 100% accurate. Anyone who has interacted with Alexa, Siri, or Google Home is aware of the great strides AI technology has made in understanding natural language. Nonetheless, there are still major limitations.

[2] I hope so; otherwise, I'd be sued for plagiarism.

from your phone last month, or the podcast you listened to this morning while jogging, the format of the underlying data was identical: a series of 1s and 0s.

However, we don't live in a digital world: sights and sounds enter our senses as waves.[3] Therefore, to be of any value to electronic devices, we must somehow convert and represent sensory input—indeed all types of data—as a series of 1s and 0s. As noted above, we call this process *digitization*.

> *Digitization* is the process of converting data into a binary representation that consists solely of 1s and 0s.

Okay, that's a definition that might be worthy of inclusion in a dictionary. But what does it mean in a practical sense?

Keyboard Characters

Let's start with a simple example. A computer keyboard transmits every character you type not as a letter but as a unique numeric code. For simplification, the computer industry has grouped and standardized these codes into *character tables*, the most common of which is ASCII (American Standard Code for Information Interchange, pronounced "AS-KEY").

For example, the ASCII representation of a lowercase "a" is the value "97"; a capital "A" has the value "65."[4] Thus, when you press the "a" key, your keyboard transmits the value "97" to the computer. Similarly, when you press the SHIFT and "a" keys together (to denote a capital "A"), your keyboard transmits the value "65." (We'll describe how that happens later in the text.)

ASCII encoding supports the displaying and printing of characters as well. For example, if the letter "A" appears on a printed page, the computer transmitted the value "65" to the printer.

Digitized Images

Next, let's look at how computers manage the graphics (images) that appear on your phones and laptops. We can gain an understanding of this processing by reviewing how humans see.

When we focus on an image, like a beautiful landscape, light rays reflecting from the objects enter our eyes through a lens called the *cornea*. The *cornea*'s primary function is to focus light before it passes through the rest of the eyeball.

The focused light eventually falls on the *retina*, which processes the incoming light using a layer of cells called *photoreceptors*. The *photoreceptors* decompose the light into a series of "points" (or "dots") based on characteristics such as color and intensity. The *retina* forwards this information to the brain (via the *optic nerve*); the brain then interprets it and creates (reconstructs) the images we "see."

---

[3] We'll ignore the senses of taste, touch, and smell.
[4] ASCII values might appear arbitrary; however, I assure you there was considerable deliberation in its design.

FIGURE 3.1   Pixels in a photograph.

Similar to the way your eye functions, we represent images in the digital world as an array of picture elements, called *pixels*, each of which represents the color and intensity of a particular "dot" of the image.[5] For example, when a digital camera captures an image (i.e., you snap a photo), it converts the light it receives through its lens into a series of *pixels* and stores them in a file (in a known format, e.g., JPEG, GIF, PNG, etc.). Later, when you view the image, display software copies the *pixels* in the image file onto your screen or directs them to a printer to create a printed photo.

Today's *pixel* encoding schemes can represent about 16 million colors; this closely correlates to what the human eye can perceive. Please refer to Figure 3.1 for an exaggerated example of *pixels* in a photograph.

Digitized Music and Sound

Now, let's consider a slightly more complicated example. I'm sure you're familiar with music CDs and some types of audio files such as MP3s and WAVs. But have you ever considered how computers process sound?

In nature, sound exists as waves that continuously change value.[6] To make sound waves suitable for processing by electronic devices, we use a component called an Analog-to-Digital Converter (ADC) that continuously samples an audio signal (e.g., music) and converts it to a numerical value.

For example, to create a digital version of a song for inclusion on a CD, an ADC samples the audio signal 44,100 times per second. That means that for every second of music, there are 44,100 individual values recorded onto the CD.

---

[5] This concept is not new. Have you ever looked closely at a picture in a newspaper? If you did, you'd see a series of dots that comprise the image.
[6] We define these types of signals as *analog*.

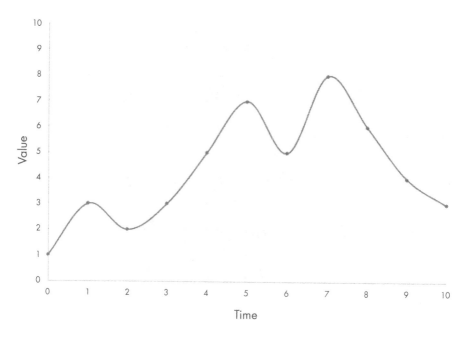

FIGURE 3.2   Sound sampling.

Figure 3.2 illustrates this process. The blue line depicts the sound wave of the song we're playing. The horizontal ("X") axis represents time; the vertical ("Y") axis represents the signal's value at a given moment. At each point in time, the ADC samples the audio signal to determine its value. Thus, at time 1, the signal's value is 3; at time 7, the value is 8.

To "play" a CD or MP3 file, the computer must reverse the process. That is, it must create an analog signal (i.e., a sound wave) from a digital (numeric) representation. To do that, it uses a Digital-to-Analog Converter (DAC) to read the sampled values and transform them into an analog signal suitable for playback through an amplifier and speakers.

In both cases (capture and playback), the key point is that electronic devices process data in its digital form. Thus, that MP3 file you emailed to a friend does not contain "waves" but rather a series of numbers.[7]

Other Examples of Digitization

The digitization examples described above are among the most common. However, the use of digitization is even more commonplace than you might think:

- Medical Instruments (e.g., automated blood pressure cuffs)

- Shazam (identifying song titles)

- Biometric Identification (e.g., facial recognition)

There are far too many to include, and the list is continually growing.

---

[7] Music files also contain other data such as "control" information for use by playback software.

Data Loss

In the examples above, we presented images that contained a specified number of *pixels* (i.e., the number of "dots") and sound files sampled at a specified interval (e.g., 44,100 times per second). In general terms, we call this the *sampling density*.

At this point, you may be wondering whether sampling densities are somehow pre-determined. The answer is no. They can vary—but there are some implications we must consider.

For example, if you take another look at Figure 3.2, you'll see that the sound wave changes value *between* the sampling points. A similar issue arises when digitizing images: photos contain "color" *between* the pixels. Thus, you may be wondering whether we lose the data between the sampling points during the digitization process. Simply put, the answer is: yes. Not surprisingly, we call this *data loss*.[8]

We can reduce *data loss* by increasing the sampling density: the more frequently you sample, the smaller the "gaps" become. There are tradeoffs, however. Increasing the sampling density also increases the size of the digital data sets, making them more costly to store (increased disk space) and transmit (longer upload/download times).

But we must also consider the limitations of human sensory reception: How much color change can the human eye perceive? How much variation in sound can the human ear detect? Thus, as a compromise, engineers develop sampling densities small enough to remain practical, yet large enough to ensure that humans usually remain unaware of the data loss (i.e., music sounds "live," pictures look "sharp").

ADVANCED TOPIC: OKAY, SO WHAT'S A "BIT"?

As we have seen, digital data is the *lingua franca* of electronic components. It's the reason why we refer to them as "digital devices." However, to fully understand how computers process digital data, you'll need a "bit" more information.

Thus far, the examples we've used represented digital data in a familiar way. That is, we've expressed values using *decimal* notation or, more precisely, the Base-10 number system that we commonly use every day. For example, we observed that keyboards represent capital "A" as the value "65."

Unfortunately, for reasons that will become clear shortly, computers cannot use Base-10. However, before discussing how electronic circuits process digital data, let's quickly review how number systems work.

When we write a number such as "456," we intuitively interpret it as represented in Table 3.1. Progressing from right to left, each column's value increases by a power of 10 (hence the name, Base-10). The right-most column (the one containing the digit "6" in the example) is the 1's column. The one to its left (the one with the value "5") is the 10's column. And the leftmost column (the one with the digit "4") is the 100's column.

---

[8] There is also another factor in the digitization process called *quantization* that can affect data loss. For our purposes, we can ignore this issue.

TABLE 3.1   Base-10 Number Representation

Base-10 Digit Value	Operation	Base-10 Column Value		Total
4	Multiplication	100	=	400
5	Multiplication	10	=	50
6	Multiplication	1	=	6
Total (addition)				456

We thus "compute" the value of "456" by multiplying each digit by its corresponding column value, then summing the intermediate results. Because we learned Base-10 at a very young age, we are no longer aware that we're performing this calculation; it's become second nature.

Something else I should point out is that the Base-10 number system requires ten digits to display all the possible values in a single column: 0, 1, 2, 3, 4, 5, 6, 7, 8, and 9. Please note, however, there is no *single* digit representing the value *ten*. The value ten requires two digits: a "1" in the 10's column followed by a "0" in the 1's column. (This is important; please keep it in mind as you read the material that follows.)

Although Base-10 is natural to us, it's not well-suited for the digital world. Instead, electronic devices use *binary numbers* or Base-2. We form *binary numbers* in the same manner as values in Base-10, but we use fewer digits (just 0 and 1), and the column values increase by powers of two, not ten.

Table 3.2 shows how to interpret the *binary number* "101." Moving from right to left, column values in Base-2 are 1, 2, 4, 8, 16, 32, 64, and so on. The basic computation is identical to the Base-10 example above; the only difference is the column values. Thus, in Base-10, "101" equals the value *one hundred and one*. Whereas in Base-2, "101" equals *five* (5).

Like Base-10, Base-2 has no single digit representing its base value "2"; it requires two digits: a "1" in the 2's column and a "0" in the 1's column. Thus, the Base-2 (binary) number written as "10" has a Base-10 value of "2." Please take a moment to convince yourself of that.

Does this seem strange? Well, another example might help clarify things. Let's convert the Base-2 (binary) value "10101" into its Base-10 equivalent. Table 3.3 illustrates the process.

In computer parlance, we refer to each binary (Base-2) digit as a *bit*, a portmanteau of the words *binary* and *digit*. Bits[9] serve as the basic unit of information in all digital

TABLE 3.2   Base-2/Binary Number Representation

Base-2 Digit Value	Operation	Base-2 Column Value		Base-10 Total
1	Multiplication	4	=	4
0	Multiplication	2	=	0
1	Multiplication	1	=	1
Total Base-10 (addition)				5

---

[9] IT Professionals commonly use other number bases as well such as Octal (Base-8) and Hexadecimal (Base-16).

TABLE 3.3   Converting Binary "10101" to Base-10

Base-2 Digit Value	Operation	Column Value		Base-10 Total
1	Multiplication	16	=	16
0	Multiplication	8	=	0
1	Multiplication	4	=	4
0	Multiplication	2	=	0
1	Multiplication	1	=	1
Total Base-10 (addition)				21

electronics. However, billions of individual bits flying around a computer would cause a world of confusion for the programmers who had to deal with them. Therefore, to simplify matters, engineers combine bits into larger groupings:

- Eight bits form a *byte*.

- Four *bytes* form a *word*.[10] Thus, there are thirty-two *bits* in a *word*.

- One *kilobyte* (usually pronounced "one-kay") is 1,024 *bytes*.

- One *megabyte* (usually pronounced "one-meg") is 1,048,576 *bytes* or 1,024 *kilobytes*.

- One *gigabyte* (usually pronounced "one-gig") is 1,073,741,824 *bytes*, 1,048,576 *kilobytes*, or 1,024 *megabytes*.

Armed with this knowledge, let's return to our keyboard example. We initially stated that the numerical representation of a capital "A" was the value "65." However, that's its human-compatible, Base-10 representation. Can you compute what its binary equivalent would be?

The answer is: "1000001." Thus, when you type SHIFT-A, your keyboard transmits a *byte* that contains the *bit* values "01000001." Note that although the representation for the letter capital "A" requires only seven bits, the leading zero is needed because, as we learned above, a *byte* comprises eight *bits*, so we "pad" the *byte* with an extra "0."

You may be asking yourself: Why do computers use the Base-2 (binary) number system and not Base-10? It's because it's very convenient to create transistor circuits that only need to recognize binary values, such a 0 and 1, on or off, yes or no, and true or false. This will become clear in upcoming chapters when we focus our attention on hardware design.

SUMMARY

Digitization is the foundation of modern information processing. Once data exists in a digital form, computers can process it in ways only limited by their programmers' imagination.

Equipped with this knowledge, we can now focus our attention on understanding how computers process digital data. We'll begin that discussion with an overview of computer hardware in Chapter 4.

---

[10] This holds true for most modern-day computers. However, there are exceptions.

What Is a Computer?

Computers are like Old Testament gods; lots of rules and no mercy.

JOSEPH CAMPBELL

OVERVIEW

Given their speed and accuracy, computers often take on an air of magic. But that is simply not the case. As we'll begin to understand in this chapter, computers are machines—nothing more, nothing less. And although they are complicated to design and build, their operation is simple in concept.

Let's dispel one myth at the outset. Despite its complexity, hardware is neither *smart* nor *intelligent.* Any apparent "magic" derives from the fact that it can execute instructions very quickly and very accurately. For example, at the time of this writing, some smartphones can perform *billions* of instructions per second. And the fastest computers in the world—called *supercomputers*—can execute more than 18 *quadrillion* instructions every second. It's mindboggling.[1]

We can divide a computer's functionality into several major categories: hardware, software, and networking.[2] We'll begin our discussions with a broad overview of each of them. However, as with any discipline, we'll need to get comfortable with some terminology.

Collectively, the term *hardware* refers to the electronic components that comprise a computing device. You may have heard of some of these: Central Processing Unit (or CPU), memory, bus, disk storage, screen, etc. When you pick up your smartphone, you hold all these components in the palm of your hand.

Nonetheless, compared to software, hardware holds a conceptual advantage: it's tangible. We can touch it. Software, however, is a tad more nebulous. It's the term we use to

---

[1] By comparison, the guidance computers used in NASA's Apollo Program—the one that landed humans on the moon—could execute less than 50 instructions per second. Yes, that's *fifty* per second. That astronauts traveled to the moon and back with such limited computing power is also mindboggling.

[2] Strictly speaking, this list is not exhaustive.

refer to the collection of instructions that the hardware executes. That is, using computer languages,[3] software developers create a series of instructions that, when executed, control the behavior of the underlying hardware.

For example, when you touch the icon representing the email app on your smartphone, the underlying software executes a series of instructions that launches the application. Once running, instructions contained within the email application itself download and display your messages.

Please take a moment to consider the power of this paradigm. The same underlying hardware that runs your email app can also manage your calls, play your music, file your taxes, etc. The point is that engineers design hardware to execute a finite set of instructions. Software programmers arrange and combine these instructions in unique ways to deliver specific functionality. Thus, only human imagination limits the types of applications computers can provide.[4]

The remaining major category of computer functionality is *networking*. This term refers to a computer's ability to interact with other systems and electronic devices.

Although you might not be aware of it, you engage in networking every day. For example, when you make a telephone call using your smartphone, it connects with the other party via the cellphone network. You might also own a desktop computer that "plugs into" a router to gain access to the Internet. If you use a wireless mouse, it connects to your laptop using Radio Frequency (RF) technology.

As I'm sure you're already aware, computers are potent tools in and of themselves. However, their power multiplies when they integrate with other devices.

CONCEPTUAL DESIGN

Although electronic circuitry has become quite complex, the major hardware components of computer systems are relatively straightforward in concept. Nonetheless, to simplify our discussions, we'll adopt a "divide and conquer" approach. Thus, throughout the remainder of this chapter, we'll decompose computer hardware into its constituent components and provide a conceptual understanding of each of them. In the chapters that follow, we'll dive into more detail.

One caveat: In the sections below, we'll discuss a "conceptual" design of a modern digital computer. However, due to the sophisticated manufacturing techniques employed today, it's not likely that you'll find any system (be it a smartphone, tablet, smartwatch, etc.) designed exactly as described below.

Hardware

If we were to remove the outer casing from a computer or smartphone, we would see such items as chips (self-contained semiconductor components), cables, disk drives, fans, and slots to hold expansion cards.[5] Figure 4.1 provides an example.

---

[3] We will discuss computer languages and programming in Chapter 7.
[4] There are practical considerations: cost, schedules, user benefit, etc.
[5] Expansion cards are self-contained electronic circuit boards that extend or enhance a computer's functionality.

FIGURE 4.1   Inside a PC.

Despite the apparent complexity, much of the hardware displayed in Figure 4.1 is not directly involved in computing, *per se*. Instead, many of these components provide supplementary services such as power, cooling, connectivity, etc. We'll ignore the ancillary hardware for the moment and focus on the *motherboard*, which hosts most of the actual computing components (see Figure 4.2).

In the sections that follow, we'll identify and clarify the *motherboard*'s essential components.

Motherboard Components

The *motherboard* serves as the foundation of electronic devices. It hosts all the components that make a computer function like, well, a computer. Figure 4.3 provides a simplified,

FIGURE 4.2   PC motherboard.

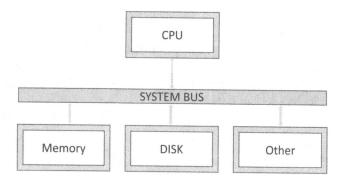

FIGURE 4.3 Motherboard components.

logical representation of a motherboard's major components. The sections that follow describe the functionality of each element.

CPU The CPU, or Central Processing Unit, serves as the "brain" of the computer in that it executes most instructions and manages the operation of other devices. A single CPU can execute one instruction at a time. However, most modern computers come equipped with multiple CPUs,[6] allowing them to execute more than one instruction concurrently.[7] We'll take a closer look at CPU design in Chapter 5.

Memory Computer memory is the storage location for all instructions and data currently executing within a computer. Before a CPU can run a program, all its instructions and data must reside in memory.[8]

In general, there are two types of memory: RAM and ROM. RAM, or Random-access Memory, is *volatile*—it loses its state when the computer loses power. Most of the programs and apps you run reside in RAM while they're executing. (Have you ever lost work because of a power failure? Now you know why.)

As its name implies, ROM, or Read-only Memory, is readable but not writeable (i.e., its value is *immutable*). It hosts data that rarely (if ever) changes during the lifetime of a device (e.g., the computer's serial number).

Disk Drives A disk drive is a device that can store electronic data permanently.[9] Disk drives come in many "flavors." Common

---

[6] You may have heard such terms as dual-core, quad-core, multi-core.

[7] Computer scientists refer to this as *multiprocessing*.

[8] We refer to this as the *Von-Neumann Architecture*. Von-Neumann proposed this design in 1945 and it still serves as the basis for most modern digital computers.

[9] Permanent does not mean eternal. Even the best disk drives fail. Back-up important data regularly.

examples include magnetic drive (hard drive), optical drive (CD and DVD), and Solid-state Drive or SSD (flash memory).

Hard drives store data magnetically on specially coated rotating platters. They typically ship in two configurations: internal (residing inside the computer's housing) and external (connecting to the computer using a cable) (see the section on *Peripheral Components* below).

Optical drives[10] operate like disk drives but use lasers to read and write data. They also support internal and external configurations.

Lasers SSDs have no moving parts. Instead, they use integrated circuits (usually flash memory) that can retain state (data) in the absence of power.

Bus In computer architecture, a *bus*[11] transports data among connected components. It allows all devices to communicate with the CPU and each other (when appropriate). In essence, it serves as the data highway for the system. For example, when the CPU needs to fetch the next instruction to execute, it issues a request to the memory component via the bus.

Peripheral Components

Computers that don't connect to any other devices are of limited use. Consider the following scenario: using your laptop and your favorite word processing package, you've just completed writing your first Pulitzer-prize-winning novel. However, how many folks would get to read it if you couldn't email it, or print it, or copy it onto a flash drive and send it to your publisher? You'd have to hand your laptop to folks and let them read your novel on its screen. Not very convenient.

Fortunately, when it comes to connecting computers to external components (called *peripheral devices* or, more simply, *peripherals*), we have many choices. You should be familiar with some of them already: mice, keyboards, and monitors. Below are some additional examples. (As an aside, I'm including this discussion in this section because most folks view peripherals as hardware devices.)

Printers Printers can be of varying technologies (inkjet and laser) and quality (credit card receipts vs. photo printers) and connect to computers using cables or wirelessly via network connections.

---

[10] We refer to such devices disc drives, as opposed to disk drives. Please note the difference in spelling.
[11] Modern computer designs typically employ multiple busses (e.g., address, data, and control).

Flash Drives Flash drives (sometimes called *thumb drives*) are portable disk drives that allow users to store and transfer files from one machine to another. They also serve as convenient backup devices.

Biometrics Now becoming mainstream, biometric devices such as fingerprint readers, retinal scanners, and facial scanners provide convenient and secure user authentication mechanisms based on physical characteristics.

This list shows a small sampling of the types of available peripheral devices. There are hundreds of others, and new products come to market almost every day.

The question you might be asking now is: how can all these different devices connect seamlessly to computers? We'll need a bit more information to answer this question and will return to it later in the text.

Software

If hardware—the CPU in particular—is the *brain* of a computer, then software is the *mind*. It's the collection of instructions and data that controls all system processing. In the absence of software, the hardware would remain idle, just wasting electricity.[12]

It's software that:

- Determines how to delete a character after you press the BACKSPACE key when using MS Word

- Computes the direction of travel when steering a car while playing Grand Theft Auto

- Calculates the effect of changing light when editing a picture with Photoshop

We'll begin our discussion on software by categorizing its types. In subsequent chapters, we'll provide examples of how to create and execute it.

System Software

System software manages the entire computer. Specifically, it controls and secures all hardware components, provides access to shared resources, and oversees program execution. It also governs access to all system resources and ensures a level playing field for all applications.

For example, suppose a system permitted individual programs to access disk drives directly. In that case, any of the following issues might arise:[13] Applications could "deadlock" contending for the same file, malicious software could gain unauthorized access to data, and every program would have to modify code whenever system administrators

---

[12] I make this comment with tongue firmly planted in cheek. Obviously, computers require both hardware and software to function. I don't want electrical engineers sending me nasty emails.

[13] These problems did arise in the salad days of computers.

upgraded to a new model. By consolidating such processing within system software, access to all resources remains controlled, uniform, and secure.

There are several subcategories of system software.

Operating System The Operating System (OS) controls all aspects of program execution. It loads all applications, enforces security, and manages access to all resources. Its tasks include:

- Allocating memory to each running process

- Terminating ill-behaving applications

- Prioritizing access to shared resources (e.g., disk drives)

 Examples of some common OSs include Microsoft Windows, Apple's macOS, and Linux (used in major software shops like Google and Amazon).

Device Drivers Device drivers serve as the "glue" between the OS and most hardware components. Consider that there are many types of disk drives on the market. They are all unique; thus, no OS could "understand" the intricacies involved to manage every one of them. To address this issue, disk manufacturers provide software modules (*device drivers*) that "know" how to interact with the OS and manage the disk.

 As an example of how device drivers function, consider the following scenario. While drafting the manuscript for this book, I periodically pressed CTL-S to save my work.[14] In response, MS Word called upon the OS, MS Windows, to copy all my changes to disk. MS Windows, in turn, passed on the request to the device driver, which wrote the data onto the drive.

 Note that both MS Word and MS Windows didn't need to know any details of the underlying disk hardware. Thus, if I replace my disk drive with a newer model, the new device driver will manage all the specifics, and there is no need to modify MS Word or MS Windows. Very slick.

Maintenance Tools Most OSs include a set of tools that help administrators and users manage the system. They typically ship as a suite of utility programs that you can execute when the need arises. For example, you may have used a program called CHKDSK to diagnose and repair disk anomalies.

---

[14] Current versions of MS Word provide a feature called AutoSave that obviates the need to do this. Nonetheless, I'm a creature of habit and still do it.

Application Software

At the risk of appearing trite, *application software* is everything else. Games, video editing packages, music programs, word processors, browsers, search engines, spreadsheets … the list is endless. That's what makes computers so powerful: the same hardware and OS platform can host an infinite variety of applications. The only limit is human imagination and need.[15]

We will discuss how software engineers design, develop, and deploy applications later in the text.

Networking

Networking is the capability by which computers connect to other computers, external devices, and the rest of the world. Where would we be without email, texting, video chats,[16] and social media?[17] The world, via our smartphones, computers, and the Internet, is connected unlike any other time in history.

At a conceptual level, there are two primary networking layers: the *transmission medium* and the *protocol*.[18] The *transmission medium* is the pathway (often called a *channel*) between communicating entities. In broad terms, we can classify *channels* into two groups: wired and wireless.

Wired media requires a physical connection between communicating devices. Some examples include traditional telephone lines, coaxial cables, and fiber-optic links. In contrast, wireless channels do not need physical connections. Data transmission occurs using infrared (light waves), radio waves, and microwaves (among others).

Riding atop the transmission channel are protocols that establish rules and conventions by which cooperating entities (e.g., computers) send and receive data. You may be aware of some of these: Transmission Control Protocol (TCP), Internet Protocol (IP), Hypertext Transfer Protocol (HTTP), and File Transfer Protocol (FTP). Together, the first two (TCP and IP) form the backbone of the Internet; we usually referred to them with the single moniker: "TCP/IP."

We can demonstrate how these layers work using old-time technology: a telephone line. Every reader of this book has used a telephone to communicate with other individuals. Most of us have also used a telephone line to send or receive a fax. In both cases, the telephone line serves as the transmission medium, and either our voices or the electronic messages exchanged by the cooperating fax machines serve as the protocol.

A more modern example would be accessing a Web page. When we type a URL into our browser (e.g., GOOGLE.COM), our PC and the Google servers exchange information over the Internet using TCP/IP and HTTP. As depicted in Figure 4.4, this request could originate from a desktop computer hard-wired to a router or wirelessly via a Wi-Fi channel. In this

---

[15] One small caveat: humans are highly creative and can envision programs that would require processing power beyond what is available at a given time.

[16] An extremely hot topic now as I'm writing this section during the COVID crisis.

[17] Maybe not for the better. Ironically, social media may be making us less social.

[18] Network engineers and software developers deal with many more layers. So will we later in the text.

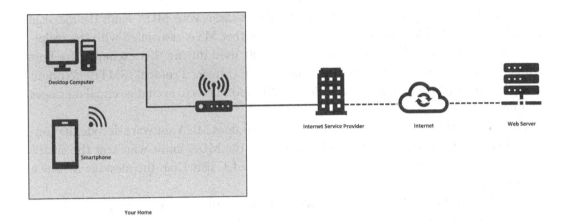

FIGURE 4.4   Web page request.

example, the protocols remain the same (TCP/IP and HTTP), but the transmission channels vary.

In short, channels move *bits*; protocols define how communicating parties interpret them.

ADVANCED TOPIC: LAYERING

As we saw in the previous example, *layering* allows us to combine channels and protocols in ways that enable parties to communicate effectively. However, layering is a technique that has uses beyond basic networking.[19]

For example, let's consider the sending and receiving of email messages. Obviously, as per above, you'll need some basic networking connectivity. In addition, programs exchanging messages need to agree on the following:

- Where does the message begin?

- Where does the message end?

- What type of content can the message contain?

- Who sent it?

- Who's the intended recipient?

I could go on, but I think you get the point. Let's take a peek at how the IT community addresses issues like this.

First, we need to establish a protocol that allows email client applications (e.g., Outlook, Gmail, Spark—generically, Mail User Agents or MUAs) to exchange messages with email servers (called Mail Transfer Agents or MTAs). The MTA's role is to route email messages from a sender's MUA to a recipient's MUA.

---

[19] As we will see throughout this book.

For example, after you compose an email and press SEND, your MUA sends the message to an MTA. The MTA forwards the message to another MTA associated with the recipient's email address.[20] The most common protocols used during this exchange are Post Office Protocol version 3 (POP3) and Simple Mail Transfer Protocol (SMTP). Without going into detail, these two protocols allow MUAs and MTAs to exchange email messages seamlessly.

But how do we interpret the payload? That is, how does MUA software decode the content of each email message? Specifically, how does the MUA know who sent the email? How does it identify any additional recipients on the CC list? Does the message require a return receipt?

To address this issue, the IT industry adopted a standard called *RFC 5322*,[21] which specifies the content and format of email messages and how MUAs should identify senders, recipients, return receipt requests, etc.

So far, so good. But what about attachments? We've all sent and received emails with photos, documents, and music files included. How is that done? The industry created another layer called the Multipurpose Internet Mail Extensions (MIME)[22] standard, which specifies how to "attach" files to email messages.

To demonstrate how some of this works, Figure 4.5 contains a simplified depiction of the layering used to construct a typical email message. For additional clarity, I also sent myself an email so that you can see the layers "at work." Figure 4.6 contains the "raw" content of that message.

To maintain privacy (mine!), I've replaced all sensitive data with generic values. Moreover, I've highlighted in gray the information I typed when generating the message; MUAs and MTAs produced everything else you see in the example in strict accordance with the requirements specified in the protocols and RFCs noted above.

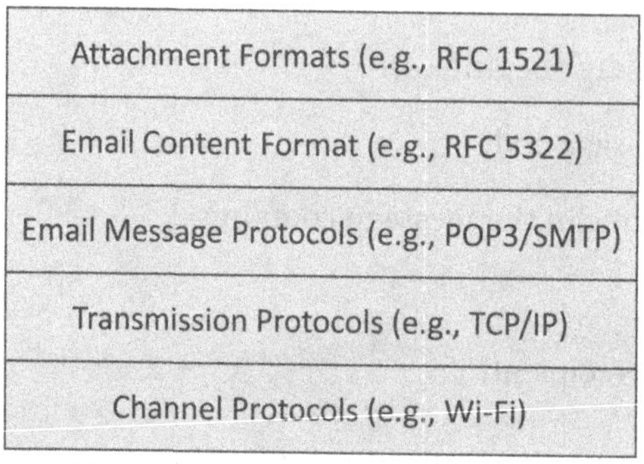

FIGURE 4.5 Simplified layering example.

---

[20] For now, we will ignore how we associate MUAs with MTAs and how MTAs route messages to each other.

[21] "RFC" stands for "Request for Comment." Distributing RFCs is the method by which IT professionals discuss and formalize new technical specifications. And for those of you scoring at home, RFC 5322 replaced RFC 2822.

[22] There are too many MIME RFCs to enumerate. The two most important are RFC 1521 and RFC 1522.

```
RETURN-PATH: <[MYEMAILADDR@MYDOMAIN.COM]>
RECEIVED: FROM [SYSTEM NAME] ([IPADDR])
          BY SMTP.[SOMEDOMAIN].COM ...
FROM: "[MYEMAILADDR]" <[MYEMAILADDR@MYDOMAIN.COM]>
TO: <[MYEMAILADDR@MYDOMAIN.COM]>
SUBJECT: THIS IS THE SUBJECT LINE
DATE: [DATE] [TIME] -0400
MESSAGE-ID: <[MSG-ID@MYDOMAIN.COM]>
MIME-VERSION: 1.0
CONTENT-TYPE: MULTIPART/ALTERNATIVE;
          BOUNDARY="----=_NextPart_000_0055_01D62E86.XXXXXX"
X-MAILER: [NAME OF MAIL PROGRAM]
THREAD-INDEX: [THREAD-INDEX-ID]==
CONTENT-LANGUAGE: EN-US
------=_NextPart_XXXX_XXXX_XXXXXXX.XXXXXXX
CONTENT-TYPE: TEXT/PLAIN; CHARSET="US-ASCII"
CONTENT-TRANSFER-ENCODING: 7BIT
TO ME,
THIS IS THE BODY OF THE EMAIL.
REGARDS,
ME
```

FIGURE 4.6   Sample email message.

Adding protocol layers is one way the IT community extends functionality as needs change. As we'll see, layering is also a generic technique that IT professionals repeatedly employ throughout all aspects of software development.

SUMMARY

This chapter introduced the three most important aspects of computer system design: hardware, software, and networking. We now have a conceptual understanding of how these pieces fit together and how they interoperate with each other.

But we've only kicked the tires. In the next series of chapters, we'll begin to peek underneath the hood.

CHAPTER 5

Internal Hardware Components

Don't explain computers to laymen. Simpler to explain sex to a virgin.

ROBERT A. HEINLEIN

INTRODUCTION

In Chapter 4, we presented a high-level description of a computer system's major components. In this chapter, we begin to "drill down" and gain a deeper understanding of how these pieces fit together.

As you begin reading through the sections below, please don't be concerned if you have some questions. Computer hardware design is one of those subjects wherein it's difficult to understand all the pieces until you comprehend the whole. I assure you that we'll tie everything together by the end of the chapter.

THE CPU

Previously, we noted that the Central Processing Unit (CPU) serves as the "brain" of a computer because it controls all aspects of system execution. However, although we often speak of it as a single element, the CPU comprises several significant subcomponents (see Figure 5.1).

In the sections that follow, we'll introduce some high-level concepts then explain each subcomponent in detail. Following that, we'll demonstrate how all the elements work in unison to run programs and execute instructions.

The Control Unit

As its name implies, the *Control Unit* (CU) oversees most of the computing operations that occur within the system. It orchestrates processing within the CPU and schedules the actions of some other components that connect to the bus (e.g., main memory). Among

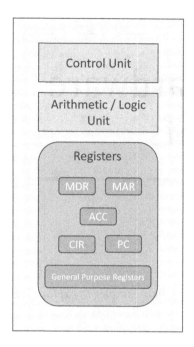

FIGURE 5.1   Major CPU subcomponents.

its many tasks, the CU ensures that the current instruction executes properly, the "next" instruction is "queued up," and that it has "fetched" all required data from memory.

Registers

Registers (or, more accurately, *hardware registers*) are high-speed memory repositories under the CPU's direct control. In modern hardware design, there are general-purpose registers that serve as "scratchpads" and special-purpose registers that perform specialized functions. (The latter group is often "hardwired" to reduce execution time.)

As it begins an execution cycle, the CPU will usually copy one or more data elements from memory into one or more *registers*. It will then perform the specified operation as indicated by the instruction and copy the results back to memory. Alternatively, the CPU may leave data "in place" so that it's available to subsequent instructions (i.e., the values may only be intermediate results).

When reviewing Figure 5.1, you may have noticed that some registers have names. That's because system designers have reserved them for specific functions. A typical CPU contains many of these; descriptions of the most common appear below.[1]

> **IR** *Instruction Register*—The IR contains the instruction currently executing. During each instruction cycle, the CPU fetches the next instruction from memory and stores it in the IR.

---
[1] Hardware manufacturers do not have to adhere to these naming conventions.

PC   *Program Counter*—Although its name might appear odd, the PC contains the memory address of the *next* instruction that the CPU will execute. Typically, this is the instruction immediately following the one in progress. However, as we'll see below, the current operation could alter the instruction sequence. When that occurs, the CPU copies the new location (address) into the PC before initiating the next instruction cycle.

AC   *Accumulator*—The Accumulator is the register into which the CPU stores the results of arithmetic and logic operations. For example, after summing the values contained in two general-purpose registers, the CPU stores the answer in the AC. At this point, the value is available for additional operations (if it's an intermediate result), or the CPU can copy its contents to memory (or both).

MAR   *Memory Address Register*—The Memory Address Register contains the address (location) of data required by the current instruction. That is, if an operation requires "fetching" data from memory, the CPU writes the location into the MAR before initiating the request.

MDR   *Memory Data Register*—The Memory Data Register contains the data that will be written to or previously copied from memory. For example, if the CPU needs a piece of data stored at location 011011, it loads that value into the MAR (see above). Then the memory controller (discussed below) fetches the value at that location and copies it into the MAR. The CPU now has a copy of the data that it can use to complete the current instruction.

The Arithmetic/Logic Unit

Most instructions execute within the *Arithmetic/Logic Unit* (ALU). As we'll describe later in the text, software developers typically write applications in "high-level" languages (e.g., Java). Before execution, we translate[2] the high-level code into a series of *primitive instructions* or *primitive operations* that the ALU executes. We'll discuss this process at length later in the text, but at a high level, this is like our keyboard example from Chapter 3. When we type, we think in terms of letters, but "under the hood," the keyboard transmits numeric codes to the computer. Similarly, software developers write programs using high-level instructions that the translation process converts to a series of low-level codes.

---

[2] We refer to the translation process as *compilation*. We'll return to this topic in Chapter 11.

There are several classes of instructions that run within an ALU; the following sections provide some examples. However, before we jump in, I'd like to make two comments: first, the primitive operations described below are examples and are not representative of any actual operations included in any computer's instruction set currently in production. Also, please keep in mind that the ALU uses binary (Base-2) arithmetic[3] for all its computations.

Arithmetic Operations

As you might expect, computers are "whizzes" at math. Below are a few examples.

ADD v1, v2   Add v1 and v2 and place the result in a register.
SUB v1, v2   Subtract v1 from v2 and place the result in a register
INCR v1   Add 1 to the value v1 and place the result in a register
DECR v1   Subtract 1 from the value v1 and place the result in a *register*

Logic Operations

Computers are nothing if not logical—sometimes to the point of frustration. Nonetheless, although it may seem counterintuitive, instructions don't always execute in strict sequence. On the contrary, developers build "decision processing" into their programs to execute specific instructions in response to particular events.

For example, consider a dialog box that pops up "asking" whether you want to save your work before you exit a program. If you press YES, then one set of instructions executes; if you select NO, then another section of code runs.

To implement such processing, programmers use *logical operators.*[4] In contrast to the *arithmetic operators* discussed above, *logical operators* don't compute values as such. Instead, they determine whether a set of one or more *conditions* are TRUE or FALSE.

As an example, please recall that for a keyboard to transmit a capital "A," the user must press the SHIFT *and* A keys at the same time. This is an example of an *and* condition. The *condition* is TRUE when the user presses both the SHIFT *and* the "A" keys simultaneously; otherwise, it's FALSE. Thus, when the *condition* is TRUE, the keyboard transmits an uppercase "A;" otherwise, it sends a lowercase "A."

We can extend this idea to create arbitrarily complex *logical expressions.* For example, consider a customer at an automated teller machine (ATM) who wants to transfer money from a checking account to a savings account. The bank's software would permit this transaction given that all the following conditions are TRUE:

- The customer is the account holder of record on the checking account, AND

- The customer has permission to deposit money into the savings account, AND

---

[3] As described in Chapter 3.
[4] In his 1947 book entitled *The Mathematical Analysis of Logic*, George Boole formalized these concepts which we now call Boolean Algebra or Boolean Logic.

- There are ample funds available in the checking account to cover the amount of the transfer.

The logical AND operator ensures *all* conditions are TRUE. However, there are occasions when only *one* of several conditions must be met for some code to execute. Consider that the bank in our previous example may grant third-party signatories access to accounts. If so, we can reformulate the first condition to read as follows:

- The customer is the account holder of record OR is an authorized signatory.

Given this rule change, the banking software will allow the transaction if the customer is either the account holder OR an authorized signatory. We can now modify the conditions as follows:

- The customer is the account holder of record OR an authorized signatory on the account, AND

- The customer has permission to deposit money into the savings account, AND

- There are ample available funds in the checking account to cover the amount of the transfer.

As the previous example demonstrates, programmers may combine *logical operators* to create arbitrarily complex expressions. However, the two operators discussed above are not sufficient to address every situation. The list below contains some other commonly used *logical operators*.

NOT   This operator yields TRUE when the condition is FALSE. For example, instead of testing whether the checking account has sufficient funds (as we did above), the code could verify that the checking account would NOT become negative if it allowed the transaction.

XOR   Exclusive "OR." With a basic OR expression, condition A can be TRUE, or condition B can be TRUE, or both conditions can be TRUE. In other words, if *any* of the conditions are TRUE, the resulting OR expression yields a value of TRUE.

However, when using the XOR operator, an expression is TRUE only when either condition A is true *or* condition B is true, but *not* both. That is, if both conditions are TRUE, or if both conditions are false, the XOR operator yields FALSE.

As an example, consider a light fixture connected using three-way switches. The bulb will illuminate only when one switch is in the "on" position, but not both.

EQ A test for equality. This operator yields TRUE when two (or more) values are equal. For example, if your checking account balance *equals* zero, then the bank's software will "bounce" any checks you attempt to cash.

NEQ The **Not EQ**ual operator. NEQ functions like EQ but yields TRUE when the values it's comparing are *not* equal. For example, when you change passwords, most security software will only allow you to save the new one if it's *not equal* to your current one.

We can summarize the values of logical operations in Truth Tables. See the examples below (Tables 5.1 and 5.2).

Data Transfer Operations
CPUs spend a lot of time repositioning data. For example, they move:

- Data from disk into memory

- Data from memory into the CPU

- Computational results from the CPU back into memory and

- Data from memory to disk

TABLE 5.1   Example Truth Table for Binary Operators

Conditions		Operators		
A	**B**	**AND**	**OR**	**XOR**
T	T	T	T	F
T	F	F	T	T
F	T	F	T	T
F	F	F	F	F

TABLE 5.2   Logical True/False Operators

Condition	Operator	
T/F	**EQ**	**NEQ**
T	T	F
F	F	T

As a result, all computers include primitive operations that move data efficiently. Some examples include:

MOV r1,r2   Move the data in *register 2* into *register 1*.
LOAD r5,01100101   Move data at memory location 01100101 into *register 5*.
STOR 01100101,r7   Move the data in *register 7* to memory location 01100101.

Again, don't be concerned if the values and terminology appear strange. We'll explain everything in more detail shortly.

Other Operations

There are several other classes of operations that a typical CPU supports. Some examples appear below.

Bitwise Operations   Bitwise instructions operate on individual *bits*. Considering that we have operators that manipulate entire *bytes* and *words*,[5] bitwise instructions may seem too tedious even for the world of computers. However, there are cases when the value of an individual *bit* is significant; consider the following scenario.

When a peripheral device, such as a disk drive, completes an operation, it must have some ways to "signal" the CPU. One way that system designers implement this process is with an *Interrupt Control Register* (ICR).[6] Each bit in the ICR corresponds to one (and only one) device. When a peripheral needs the CPU, it raises "an interrupt" by setting its corresponding bit to 1.[7] (Otherwise, the bit remains fixed at 0.)

Bitwise operators allow peripherals to manipulate their specific *bit* within the *interrupt control register* without affecting any other *bits* (or their corresponding devices).[8]

Control Flow Operations   As discussed earlier, there are times when, based on a set of conditions, program execution must continue at a different location. *Control flow* operators support such "detours" in process flow.

To see how this works, let's review the processing logic in our ATM example from above.

- The program begins the transfer request by executing all the logical operations to determine whether the transaction is valid.

- If the result is TRUE, it continues execution at a location that contains code that will complete the transfer operation.

- Otherwise, if one or more condition evaluates to FALSE, the program will execute code that informs the user the system cannot complete the request.

---

[5] Please recall from Chapter 3 that both *bytes* and *words* are a grouping of *bits*.
[6] This is another example of a *specialized register*.
[7] In general, hardware engineers refer to this process as *signal handling* or *interrupt processing*.
[8] The term used to describe this type of operation is *masking*.

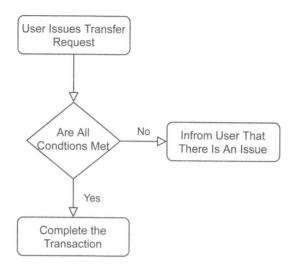

FIGURE 5.2   Example Flowchart.

We can represent this logic using a Flowchart. See Figure 5.2 for an example.

The judicious combination of logical operators and control flow constructs provides the illusion that software can respond to our every whim.

Floating-Point Arithmetic   In computers, floating-point computations are mathematical expressions containing *real* numbers (i.e., values that include a decimal point, e.g., 123.45[9]). They are among the most complex and time-consuming operations performed by the hardware.

To compensate, system engineers often equip computers with a floating-point accelerator chip (also called a *floating-point unit*, FPU, or *math coprocessor*) that contains custom circuitry specifically designed to execute floating-point operations efficiently. In such cases, the CPU directs all floating-point computations to the FPU. (You may recall that at the beginning of this section, I stated that the ALU executes most instructions—but not all; this is one example.)

Surprisingly, when it comes to floating-point arithmetic, computers are not as precise as you might think. To those who are new to computers, this is often a startling realization. It also confounds many budding programmers who cannot interpret the "weird" results they sometimes receive when their code executes.

There are several reasons for this:

- Real numbers often don't have an exact representation. Consider the value 1/3. Its decimal equivalent is .333333... *ad infinitum.*

- Remember, computers perform all arithmetic operations in binary. We lose precision when converting from Base-10 to Base-2 and back.

---

[9] In binary arithmetic, we call this a *binary point* (e.g., 101101.0010).

- It's challenging to represent real numbers in bounded spaces. All computer components (registers, memory locations, etc.) are of a fixed size. As a result, it's difficult to represent values precisely when numbers grow very large or very small.

Although most computers are sufficiently accurate for day-to-day use, this is a concern in many fields of study that require high degrees of precision. Although beyond this book's scope, computer scientists and electrical engineers have developed ways to mitigate—but not resolve—this issue.

Assembler Mnemonics and Opcodes

Modern CPUs support hundreds of primitive instructions; the few presented in the preceding sections are but a small sample. Nonetheless, some readers might have a question at this point: given that computers can only process digital values, how is it that the instructions can have names like ADD, SUB, XOR, etc.? After all, they don't look like binary numbers.

The answer is that these names—called *assembler instructions*—are for human consumption only. Collectively, they comprise what's called the computer's *assembly language* instruction set. Each *assembler instruction* (or *mnemonic*) corresponds to a unique binary value called an *opcode*. Collectively, the suite of *opcodes* forms the *machine language* of the computer. There's a one-to-one correspondence between human-readable *assembler instructions* and machine-executable *opcodes*.

Thus, software developers can write programs using an assembly language then convert the instructions into the corresponding set of opcodes using a system-supplied utility program called an *assembler*.[10] We refer to the conversion process as *assembling code*.

In summary, one way to write a computer program is as follows:

- Create a set of assembler instructions that implement the desired functionality

- Save the instructions in a file

- Use the system-supplied *assembler program*[11] to convert the assembler instructions into their corresponding machine-level opcodes, saving the output in another file. (We usually call such files *executables*. Under MS Windows, executable files have a .EXE extension as part of their name; on macOS, such files have either a .DMG or .APP extension.)

- Launch (execute) the program

On a practical note, other than a few system-level programs, developers rarely write applications using assembler mnemonics these days. It's far too tedious. It's much more likely that developers will use languages such as Java, HTML, Python, C++, etc. We'll return to this discussion later in the text.

---

[10] Computer scientists are nothing if not clever with their naming conventions.
[11] The *assembler program* does a lot more work than described here.

MEMORY

In the prior section, we discussed some memory locations (i.e., *registers*) under the CPU's direct control. In this section, we'll review a subsystem that also stores data: *main memory*.

Main memory is the component that maintains—and makes available—all the information required by the CPU to manage the system. For any program to execute, all its instructions and data must reside in memory.[12]

Conceptually, you can think of memory as rows and rows of uniquely numbered (addressed) lockers in a health club's changing room. The CPU can look inside any locker (i.e., read its contents) or replace its contents with new "stuff" (i.e., overwrite its current value with a new one).

Figure 5.3 depicts a conceptual view of memory in a computer. Each box (or *cell*) is a uniquely addressable memory location (i.e., "locker") and represents one byte of storage. (Recall that a *byte* contains 8 *bits*.) For convenience and logical representation, I've arranged the cells in rows and columns.

The numbers appearing to the left of the cells are *addresses*. Specifically, each value indicates the address of the first cell in each row (up to one megabyte in this example). I included both decimal (Base-10) and binary (Base-2) representations of each address for your convenience.

As you scan across each row, you add "1" to compute each successive cell's address. For example, the address of the cell labeled "x" is $32_{Base-10}$, the address of the cell labeled "y" is $66_{Base-10}$, and the address of the cell labeled "z" is $127_{Base-10}$. Please take a moment to get comfortable with this addressing notation before reading on.

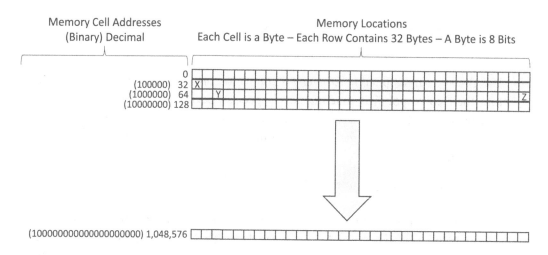

FIGURE 5.3 Main memory.

---

[12] This statement is not 100% accurate. To minimize memory usage, some program information may remain on disk until the CPU needs it.

There are several points about main memory worthy of note. First, it's *dynamic* and *volatile*. That means that programs can change the contents of memory cells at will,[13] and its contents will disappear if the system loses power or restarts.

Second, applications and system components don't have direct access to memory locations. Instead, they interact with a *controller* whose job is to store and fetch data from memory.

To see how this works, let's return to our example of *registers*. If the CPU needed to fetch the value from the cell labeled "Y" in Figure 5.3, it would insert the value 66 (actually, its binary equivalent "1000010") into the MAR and then signal the memory controller to "go fetch." After acquiring the data, the memory controller would store the cell's value into the MDR, where it would become available for use by the CPU.

THE BUS

Often called a "data highway," a *bus* is a subcomponent that facilitates data exchange among components within a computer system. There are many types and classes of busses. For our purposes, however, we need to discuss only two of them.

Internal Bus

The most common type of bus is *internal*. It serves as the pathway for data transfer among the computer's internal components. Figure 5.4 provides an example.

As depicted in the diagram, an internal *bus* consists of three sub-busses.

Address Bus	The *Address* Bus contains the address of the data that is to be read or written.
Data Bus	The data travels on this bus to/from communicating components.

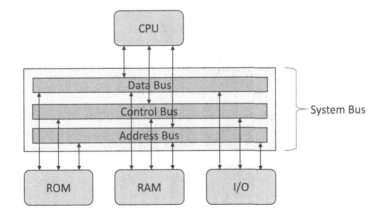

FIGURE 5.4   Internal bus architecture.

---

[13] For security reasons, programs can only modify memory locations that they "own." We'll cover this issue in more detail when we discuss operating systems in Chapter 9.

Control Bus Among its other uses, this bus serves as a control channel specifying commands such as whether the value placed on the Data Bus should be read or written at the address indicated on the *Address Bus*.

Let's take a moment to understand how a *bus* facilitates data transfer and component interconnectivity.

Assume that after computing the sum of two numbers, the CPU needs to store the result (let's say, 011011) at a specific memory location (e.g., 011001001).

To initiate this operation, the CPU would:

- Indicate on the Control Bus that this is a WRITE operation

- Copy the computational result (011011) onto the Data Bus and

- Set the memory location on the Address Bus (011001001)

At this point, the data transfer is complete.

One last point: Components don't connect directly to the bus. As depicted in Figure 5.5, there is typically a *controller*[14] that "sits in-between" and serves as the interface between the bus and the device's data.

In practice, bus designs vary from system to system. In addition, they can be unique to a manufacturer or may leverage standard industry designs. Modern systems often use a combination of both.

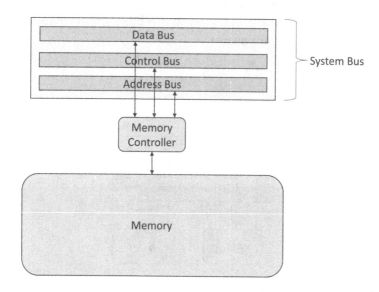

FIGURE 5.5   Memory controller.

---

[14] Controllers are "intelligent" devices driven by a type of software called *firmware*.

FIGURE 5.6   External USB hub.

External Bus

Anyone who has connected a peripheral device (e.g., a printer) to a computer using a USB cable has already used an *external bus*. As an acronym, USB stands for Universal Serial Bus. Developed in the early 1990s, its design supports "plug-and-play" installation of external devices.

Today, most computers include one or more integral USB ports.[15] Users can also purchase and install a USB Hub to increase the number of devices connected to a system (Figure 5.6).

DATA STORAGE

We discussed storage devices briefly in Chapter 4. In this section, we'll take a closer look at this technology.

It's likely that the most common peripheral devices used by readers of this book are disk drives. Often called *drive storage*, these devices read and write data encoded magnetically or optically on rotating platters. Drives range in size (the amount of data they can hold), performance (the speed at which they can read and write data), and form factor (internal or external).

One note: This chapter focuses on the internal components of computers. However, because some of the storage devices discussed below support both internal and external configurations, we'll review both form factors together.

Hard Drives

Hard drives are the most popular storage media included in general-purpose computers. Most desktop and laptop PCs ship with at least one internal disk drive. Many users also purchase additional external devices for added storage and backup.

Disk drives comprise a "stack" of rotating "platters"—each coated with a thin film of magnetic material—that rotate at speeds varying from 4,200 RPMs in smaller portable devices to 15,000 RPMs in high-performance systems. As the platters spin, data (*bits*) are

---

[15] This may run counter to your experience with computers at work. For security reasons, many organizations purchase computers that don't have any USB ports or external disk drives.

FIGURE 5.7   Internal hard drive.

read and written from/to the platters by devices called *read-write heads* that "float" just above the magnetic surface.[16] Because the platters are "stacked," there are read-write heads that service both the top and bottom of each. Figure 5.7 contains a picture of a typical internal hard drive.

Disk drives group (organize) data into "tracks" on each platter. To retrieve data at a given location—like memory, each byte on a disk has a unique address—the disk controller moves the read-write heads to the appropriate track and waits until the desired location rotates under them.

Optical Drives

Optical disc drives use laser light to store and retrieve data, employing the same underlying technology used in CDs and DVDs. Figure 5.8 contains a picture of a typical optical drive.

Although you can still order them, most manufacturers no longer preconfigure PCs with "onboard" optical drives. Their usefulness has diminished due to the emergence of flash and cloud technologies (see below).

Flash Drives

A USB flash drive is a portable data storage device containing two main subcomponents: flash memory and a USB interface (see Figure 5.9 for an example).

The USB interface allows the drive to communicate with a computer. Flash memory is a portable data storage medium that maintains its state in the absence of power. (You may have also used flash memory cards in cameras, smartphones, printers, etc.)

---

[16] In the range of tens of *nanometers* above the platters.

FIGURE 5.8   Internal optical drive.

USB Flash Drives have (almost) rendered optical drives obsolete because they're smaller, faster, cheaper, and more reliable (they have exactly zero moving parts). At the time of this writing, you could purchase a ¼ terabyte (256 gigabytes) flash drive for less than $40.00.

Solid-state Drives

While USB Flash Drives are for external use, Solid-state Drives (SSDs) are their internal counterpart. Both use non-volatile memory to maintain their state in the absence of power. However, the SSD's form factor (design) allows it to replace internal hard drives.

Although they offer many advantages compared to their hard-drive counterparts (performance, reliability, lower power consumption), they suffer one drawback: they are more expensive than traditional platter-based disk technology. Nonetheless, many manufacturers

FIGURE 5.9   USB flash drive.

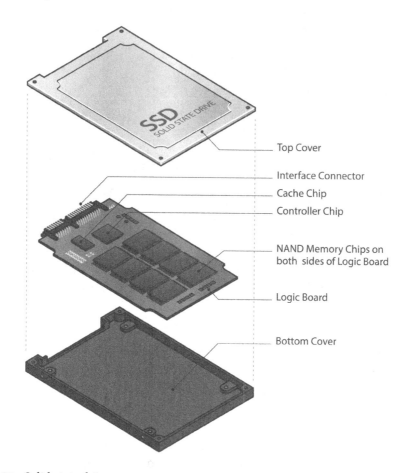

Top Cover

Interface Connector

Cache Chip

Controller Chip

NAND Memory Chips on
both sides of Logic Board

Logic Board

Bottom Cover

FIGURE 5.10   Solid-state drive.

now offer SSD options for users who don't mind paying a bit more for the increased speed
and reliability (Figure 5.10).

Cloud Drives

A Cloud Drive is remote storage that someone else manages for you, but it appears as internal storage on your computer. Once configured, you use cloud drives as if they reside inside
your computer (see Figure 5.11 for an example).

When provided by reputable vendors, cloud storage offers several benefits: it's secure,
inexpensive, well maintained, backed up, recoverable, and available anywhere there's an
Internet connection.

However, like every product, it does have some disadvantages, including increased
dependence on your network connection and (potentially) too much reliance on a third-
party vendor.[17] Nonetheless, for home users, cloud storage can provide a reliable and inexpensive backup mechanism for important data.[18]

---

[17] We refer to such dependence as *vendor lock-in.*

[18] Despite the added security features of cloud drives, I strongly urge you to encrypt important data (e.g., personal information) before storing it remotely. We'll both sleep better at night.

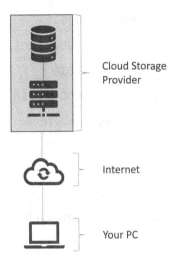

FIGURE 5.11   Cloud storage.

INSTRUCTION EXECUTION

It's time to complete the circle and understand how all the knowledge we've just gained comes together. As mentioned in an earlier chapter, a computer is a machine; nothing more, nothing less. Thus, the *raison d'être* of a CPU is as follows:

1. Fetch the next instruction (Instruction Fetch Stage)

2. Decode the instruction (Instruction Decode Stage)

3. Execute the instruction (Instruction Execution Stage)

4. Rinse and repeat

The process just described constitutes one *instruction cycle*. Admittedly, it's a simplified description—but only in terms of the details. So long as it has power, that's what a CPU does.

To take a closer look at how this works, let's consider the following statement:

ADD M1,M1

The mnemonic, ADD, is the assembler instruction; M1 and M2 represent the memory locations of the two values (operands) we want to sum.

Let's further assume the following (unless otherwise indicated, all values are in Binary or Base-2):

1. The instruction, ADD, resides at the memory location (address) 01111111

2. The opcode for the ADD instruction is 01110111 (this is the "value" that resides at memory location 01111111)

3. The operand, M1, resides at address 01000001

4. The value at location M1 is 1_{10} (Base-10) or 00000001_2 (Binary or Base-2)

5. The operand, M2, resides at address 01100010

6. The value at location M2 is 2_{10} (Base-10) or 00000010_2 (Binary or Base-2)

7. The AND instruction requires that its operands reside in registers R1 and R2 and

8. The result of our ADD instruction is obviously 3_{10} (Base-10) or 00000011_2 (Binary or Base-2)

Given the above, let's step through an instruction cycle and understand how it works. While reading through this example, you might find it helpful to refer to Figure 5.1.

1. *Instruction Fetch Stage*

   a. You may recall from our discussion of *registers*, the Program Counter (PC) always contains the address of the *next* instruction to execute. In our example, that's 01111111 (assumption 1 above). The instruction cycle begins when the CU copies the PC's contents into the Memory Address Register (MAR).

   b. In preparation for the *next* instruction cycle, the Control Unit (CU) increments the PC to point to the next instruction in sequence. This step is the reason why the PC *always* contains the address of the *next* instruction. (However, please recall from our discussions of *control flow operators* that if this were a *conditional instruction*—instead of ADD—the PC's value might change before the beginning of the next cycle. We can ignore this point for now.)

   c. Via the bus, the Memory Controller acquires the address contained in the MAR. Keep in mind that this is the *address* (memory location) of the ADD instruction (01111111), not the instruction proper.

   d. The Memory Controller fetches the instruction (opcode) and the two operands (M1 and M2) from memory and again uses the bus to copy these values into the Memory Data Register (MDR). This is the value 01110111-01000001-01100010 (per assumptions 2, 3, and 5 above). Please note that the dashes are not part of the instruction and are not present in the IR; I inserted them here solely to aid readability.

   e. The CU copies the opcode (01110111) residing in the MDR into the Instruction Register (IR).

The CPU is now ready to move onto the next stage.

2. *Instruction Decode Stage*

   a. At the start of this stage, the CPU "decodes" the instruction. As noted in 1d, after the fetch from memory, the IR contains the following bits:

   01110111-01000001-01100010

The first group of bits (01110111) is the *opcode* of the ADD instruction. Immediately following that are the *addresses* of the two operands, M1(01000001) and M2 (01100010), respectively. Thus, during this decoding step, the CPU *parses* the instruction into its constituent components: opcode (01110111), address 1 (01000001), address 2 (01100010).

b. Just like the processing used to fetch the instruction, the Memory Controller now acquires the data stored at locations 01000001 (M1) and 01100010 (M2) and, using the bus, transfers the corresponding values (00000001 and 00000010) into registers R1 and R2, respectively (assumption 7 above).

Okay, let's take a breath and summarize what has just happened. At this point, the IR contains the opcode (01110111) of the ADD instruction (as per 1d), register R1 contains the value (00000001) that resided at memory location M1 (01000001) (as per 2b), and register R2 contains the value (00000010) that resided at memory location M2 (01100010) (also as per 2b).

The CPU is now ready to enter the Instruction Execution Stage.

3. *Instruction Execution Stage*

a. To begin this stage, the CU copies the contents of registers R1 and R2 into the Arithmetic/Logic Unit (ALU).

b. The ALU performs the addition and stores the result (00000011) in the AC (Accumulator).

c. The result remains in the AC to serve as an operand in subsequent instructions (i.e., the ADD instruction in our example might be part of a more extensive computation) or until a subsequent instruction directs the CPU to copy the value to memory.

4. *Repeat*

a. *Repeat ad infinitum.* Return to step one to execute the next instruction whose address is already in the PC.

Well, that's it—a complete instruction cycle. Although somewhat tedious, there's no magic involved—just execution logic. You might want to step through this example a few times until you feel comfortable with it. Also, as you reflect on this processing, keep in mind that your smartphone or laptop can execute millions, if not billions, of similar instructions every *second*.

ADVANCED TOPIC: ARITHMETIC SHIFT INSTRUCTIONS

Sometimes computers don't execute instructions in the most obvious manner. For example, in most systems, multiplication is an "expensive" operation. However, *bit shift* operations (i.e., sliding bits left and right within a *byte* or *word*) are not. Thus, computer designers realized that because the underlying values are binary (Base-2), they could replace expensive mathematical computations (e.g., multiplying or dividing by powers of 2) with faster bit shift operations.

To demonstrate how this works, let's start with a simple example: 8 times 2. The answer is obviously 16. As a byte, the value 8 has a binary representation of 00001000.[19] The binary result, 16, is 00010000. When looking at the before and after bit patterns, it becomes clear that the net effect is that the "1" bit "shifted" one location to the left.

Now, let's multiply 8 by 4. The answer, 32, has a binary representation of 00100000. In this case, the "1" bit "shifted" left two positions.

In the first example, we multiplied 8 by the value 2. In exponential notation, 2 is equivalent to 2^1 (2 raised to the power 1). In the second example, we multiplied 8 by 4. In exponential notation, 4 is equal to 2^2 (2 raised to the power 2). Notice that in each multiplication operation, we shifted the "1" bit the number of locations indicated by the value of the exponent. (Please take a moment to get comfortable with this before reading on.)

However, the above examples don't highlight the fact that the CPU shifts *all* bits in a byte, not just the "1" bits. The multiplicand, "8," used in both computations, contains only a single "1" bit, so it's difficult to see that all the "0" bits shifted as well.

To demonstrate that the CPU shifts all bits, we need a more complicated example: let's multiply 85 by 2.

The decimal number 85 has a binary representation of 01010101. When we multiply that value by 2, the result, 170, has a binary value of 10101010. It's clear in this case that the "0" bits shifted as well. Also, please note that the CPU "padded" the rightmost bit with a "0."

Now, let's reverse the process. If we begin with 8 (0001000) and divide by 2, we compute an answer of 4 (0000100). In this case, the bits shift to the right, and the "0" padding takes place on the left.

The above examples omitted some practical considerations, such as what happens when a "1" bit "falls off" one of the edges during a shift. We'll leave such concerns to system designers.

SUMMARY

This chapter discussed the internal components of a computer: Central Processing Unit, memory, bus, and storage. We also examined the CPU's composition and learned that it contains a CU, an ALU, and temporary memory locations called Registers. We also discovered that the Bus comprises three subcomponents: the Address Bus, the Control Bus, and the Data Bus, and we described how they work in unison to transfer data among attached components. Finally, to "put it all together," we reviewed how an instruction executes using all these components.

---

[19] Remember that a *byte* usually contains 8 *bits*, that is why we need all the "leading zeros."

CHAPTER 6

Hardware—External Components

Hardware: the parts of a computer that can be kicked.

JEFF PESIS

INTRODUCTION

In Chapter 5, we focused on the internal components of computer systems. One of the recurring themes of that discussion was that CPUs would be of little value if they couldn't communicate with the other devices attached to the bus. The same holds for computers in the large: if they can't interoperate with users or other devices, they are of little practical value.

This chapter will broaden our discussions to include external components that interoperate with computers via cables or wireless connections. For pedagogical convenience, we can classify these devices into several categories: input-only, output-only, and combined (supporting both input and output functionality).

INPUT-ONLY DEVICES

In this section, we'll discuss some of the most common input-only devices. I'm sure you've used or are familiar with most of them; nonetheless, there might be some surprises.

Keyboard

In Chapter 3, we discussed how computer keyboards work. As a brief reminder, when you press keys—either singly or in combination—the keyboard transmits a unique binary value to the system, and the software interprets the input as appropriate.

For example, consider using a word processor like MS Word. When you press character keys such as SHIFT-A,[1] your keyboard transmits the corresponding value (65_{10} or

---

[1] Not every keypress results in a data transmission. The SHIFT, ALT, and CTRL keys (among others) are *modifier keys* whose function is to alter the behavior of other (transmittable) keys.

$01000001_2)^2$ to your computer, and the application (MS Word) interprets the keypress as follows: *The user wants the character "A" to appear in the text at the current cursor location.* In response, MS Word adds "A" to your document and displays it at the proper position on the page (screen) in the appropriate font.

We refer to characters like "a," "A," "1," "$" as *printable characters* because we can "see" them. That is, they physically appear on screens and printed matter. But what about other keys such as ENTER, END, and HOME? When you press them, nothing displays on the screen because they act like commands or instructions.[3]

Let's consider the ENTER key (ASCII value 13_{10}, 00001101_2) for a moment. In a word processor application (e.g., MS Word), pressing ENTER usually indicates that you want to begin a new paragraph. However, when you press ENTER after typing a URL into an address bar, a browser interprets that keystroke as a request to "fetch me some results." The point is that there is a difference between a key's value and any corresponding action that might result when you press it.

Mouse

Computer mice[4] are among a class of components called *pointing devices*,[5] which we use to position the cursor on the screen and initiate actions via button presses. As you move the mouse, it tracks its motion and transmits *deltas* (i.e., relative changes in position) to the computer. This information directs the graphical user interface (GUI) subsystem[6] to move the cursor around the screen.

There are several technologies that mice use to track movement. As depicted in Figure 6.1, older mechanical mice use a ball and a combination of sensors.

FIGURE 6.1 Mechanical mouse.

---

[2] As a reminder, this notation means 65 in Decimal (Base-10) or 01000001 in Binary (Base-2). Keyboards transmit data in binary.

[3] Some nonprintable characters support functions such as networking and device management. As a result, IT professionals often refer to these as *control characters.*

[4] Both "mice" and "mouses" are acceptable spellings for the plural of computer mouse. I prefer the former.

[5] Trackballs and touchpads are other examples.

[6] The Graphical User Interface manages the screen (e.g., windows and icons). We will return to this topic later in the text.

FIGURE 6.2 Optical mouse.

Modern optical mice have no moving parts. They use Light-emitting Diodes (LEDs) or lasers that detect movement with respect to the underlying surface (see Figure 6.2).

Image Input

In Chapter 3, we described image digitization. As you may recall, sensing components detect incoming light waves and transform them into discrete digital values called *pixels*. Applications use pixel values to process and manipulate the image. Many products employ this technology, including digital cameras, video cameras, faxes, and scanners.

Audio Input

Chapter 3 also discussed how Analog-to-Digital Converters (ADCs) capture and transform sound waves into discrete digital values. ADCs can process sound from microphones and direct audio signals (e.g., input from an electric guitar).

Barcode and Quick Response Readers

Barcodes represent data as a series of parallel lines and spaces. By varying the width of the lines and the amount of space between them, they can represent a wide range of values.

Barcode readers are optical devices that scan barcodes, convert the encoded data into its binary equivalent, then transmit the result to computer applications. They come in many forms (wand, hand-held, and stationary), and most smartphones can scan them as well.

Barcodes have become commonplace. They appear on almost all consumer products (take a glance at the back of the book you're currently reading), are included on business cards to allow easy insertion of contact information into electronic address books, and serve as tickets to cultural events (see Figure 6.3).

Quick Reference codes (QR codes) are a variant of barcodes. Originally invented in Japan during the 1990s for use in the automotive industry, QR codes quickly gained popularity and spread to other commercial and marketing applications.

FIGURE 6.3   Types of barcodes.

Optical Character Recognition

Optical Character Recognition (OCR) software scans images and converts text (both printed and handwritten) into its equivalent machine encoding (i.e., the digital representation of the characters as if typed at a keyboard). OCR has been widely used as a replacement for human data entry because it can quickly and accurately digitize printed data.[7] In particular, it has become the backbone for mail processing in the US Postal service because it can rapidly interpret and sort addresses (see Figure 6.4).

Today, many tablets and smartphones ship with a stylus that allows users to draw and take handwritten notes and onboard OCR software to convert them into their equivalent digital formats.

---

[7] The process is not perfect and often requires human review.

FIGURE 6.4   OCR process.

Speech Recognition

Speech recognition is the process that converts spoken language into digital text suitable for processing by electronic devices. Formerly a "futuristic" feature of science fiction stories (e.g., "HAL" in *2001: A Space Odyssey*, or "Computer" in *Star Trek*), speech recognition has become commonplace. For example, we now verbally direct the navigation systems in our cars to "drive home," we dictate text into word processing applications, and command our smartphones to "call mom."

Biometric Devices

Like speech recognition, biometric devices have also become commonplace. We routinely use thumbprints instead of passwords to access smartphone apps, unlock our laptops using facial recognition software, and use retinal scanning instead of ID cards at secure workplace facilities.[8]

Each type of biometric device employs different sensing technologies, but the overall process is similar:

- The scanner detects the underlying physical biometric characteristics (e.g., fingerprint loops and whorls)
- Software digitizes the input from the scanner
- Other software compares the new scan to a previously recorded copy

At the time of this writing, the most common biometric devices include:

Iris & Retina Scanners	Every human being has unique iris and retinal patterns. Iris scanners detect and digitize these differences.
Fingerprint Scanners	This is the most economical biometric scanning device. However, to counter spoofing, high-end fingerprint scanners now capture and use 3D images.

---

[8] Biometric identifiers are far more secure than passwords but are still subject to "spoofing."

Facial Recognition   Everyone's face is unique. Even identical twins have minute differences that the human eye cannot detect. It may be something as small as the relative positioning of eyebrows, the distance between eyes, or the breadth of the nose. Facial recognition software calculates "landmarks," such as the corresponding shape, size, and location of the eyes, jaw, and nose. Advanced devices also capture 3D imaging to determine the shape of the chin or nose.

Voice Recognition   Though it is often difficult for humans to discern, every individual possesses a unique voice pattern. To uniquely identify individuals, sophisticated voice recognition software can process, detect, and digitize those differences.

OUTPUT-ONLY DEVICES

Information is only useful when it's made available to its consumers.[9] This section will discuss several types of output-only devices and provide examples of how they present data.

Screens, Monitors, and Projectors

Computer screens, monitors, and projectors display information using text and graphics. They're so commonplace today that we don't give these devices a second thought. However, in the early days of computing, software developers worked on "terminals" that didn't have screens; instead, all output took the form of printed characters on rolled paper (see figure 6.5).

FIGURE 6.5   Teletype terminal.

(Editorial credit: Aitor Serra Martin/Shutterstock.com.)

---

[9] In general, the term, consumer, refers to both human users as well as other computer systems.

FIGURE 6.6   Example cast icon.

Please keep in mind that despite the sophisticated electronics packaged into modern display devices, it's still incumbent upon the underlying application software (as developed by humans) to present data effectively. Formats include text (e.g., documents and music playlists), images (e.g., photos and videos), and graphics (e.g., pie charts and scatter diagrams).

Casting

Casting is a way to project an image appearing on a small display (e.g., a smartphone's screen) onto a device with a larger format such as a TV or computer monitor. Using products like smart TVs, Google's Chromecast, Apple's AirPlay, or Amazon's Fire TV, users can "cast" (project) the image on their phones or tablets onto a large screen.

Any application that displays an icon like the one depicted in Figure 6.6 can project its image to a cooperating device.

Augmented Reality and Virtual Reality

As display technology grew more sophisticated, data presentation began to take on forms other than simple text. For example, we can display information as images, charts, icons, videos, etc. Indeed, technology has advanced to the point that we can Augment Reality (AR) or create a Virtual Reality (VR).

Using AR devices, users can enhance their interaction with the "real world." For example, someone traveling abroad could point their smartphone at a street sign, and onboard AR software would not only display the image but provide a translation of the text as well. Military aircrafts (and many mainstream automobiles) provide heads-up displays (graphics projected directly onto windshields) that allow pilots and drivers to see reality overlaid with additional data such as speed, fuel reserves, etc. (see Figure 6.7).

Alternatively, VR is a simulation of reality. VR devices and software create virtual worlds allowing users to experience anything that other humans can conceive. For example, pilots can learn how to fly new aircraft in simulators before ever setting foot in the cockpit; medical researchers can "place" patients in specific virtual environments to trigger and measure reactions; and, of course, gamers can become immersed in a virtual world (see Figure 6.8).

No longer limited to the senses of sight and sound, VR users can experience other sensory inputs such as haptic (touch), olfactory (sense of smell), and somatosensory (pressure sensations felt anywhere on the body). (See Figure 6.9.)

FIGURE 6.7   Augmented reality example.

FIGURE 6.8   Virtual reality.

FIGURE 6.9　The full VR experience.

Printers

Though considered "old school" now, printers display text and graphics on paper (as well as other materials). There are many types of printers: document, photo-quality, barcode, and plotters (see Figure 6.10).[10] They also employ several technologies (e.g., inkjet and laser).

Due to the emergence of high-resolution graphics displays on smartphones and laptops, the standardization of formats (e.g., MPEG, GIF, PDF), and the increased use and acceptance of electronic signatures, the need for hard copies—and by extension, harvesting trees for paper—is dwindling.

3D Printers

Though a bit of a misnomer, 3D printers fabricate objects from 3D digital models. Since their inception in the early 1990s, the use of 3D printers has grown from creating simple prototypes to producing precision production components. 3D printer technology has progressed to the point that designers use them to construct objects that would be difficult—if not impossible—to build by conventional means (see Figure 6.11).

Audio and Video Streaming

Most readers of this book have likely "streamed" audio and video clips. Examples include listening to a radio program on your smartphone or watching a movie on your laptop. The process is like that described in Chapter 3, except that the digital data source (i.e., an MP3 file) does not reside on your computer or smartphone but rather on a remote server located somewhere on the Internet.

---

[10] There is a technical difference between printers and plotters: printers "display" images; plotters "draw" them.

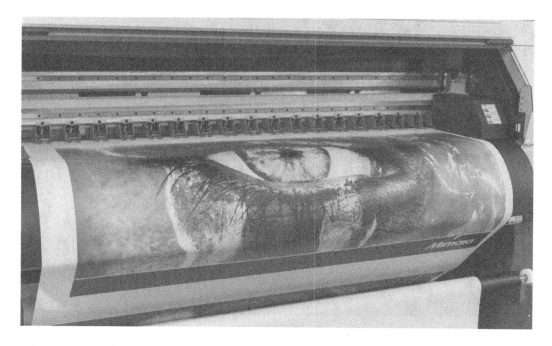

FIGURE 6.10   A plotter in action.

(Editorial credit: nikshor/Shutterstock.com.)

In most cases, streaming is as easy as using an app like Netflix or directing your browser to visit a website that offers such content. Once connected (and optionally authenticated), the webserver redirects the app or browser to another host, called a *streaming server*, which continuously transmits the video or audio content in small "chunks" called *packets*. The app or browser, in turn, forwards the packets to a "player" that decodes the inbound data

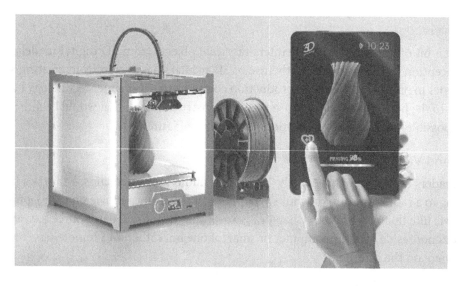

FIGURE 6.11   A 3D printer in action.

and "plays" the content as it arrives. Although it seems easy to users, streaming requires the seamless integration of many individual components and several technologies and protocols.

Speech Synthesis

As you may recall from Chapter 3, we discussed how a Digital-to-Analog Converter (DAC) transforms digitized audio data into waves suitable for playing through speakers or headphones. Though more complicated, synthesized speech is a variation of that process. In this case, however, plain text (like the words you're reading) serves as the source for the conversion.

INPUT AND OUTPUT DEVICES

Many peripherals offer both input and output capabilities. For example, we've already discussed disk and flash drives. The sections below discuss additional examples of such devices.

Touchscreens

Touchscreens are commonplace today. Most users of smartphones use them intuitively without any training: the screen "paints" (i.e., displays) a set of controls (e.g., "buttons" and "menus"), and users touch the screen to make their selections.

Smart Speakers

Who among us is not familiar with smart speakers such as Alexa and Google Home? These devices function by listening for a triggering phrase[11] (e.g., "Alexa" or "Hey Google"), then converting subsequent spoken words into text so that it becomes actionable.

For example, many folks start their day by asking Alexa or Google Home, "What's today's weather?" In response, these devices convert the spoken request into text, interpret its meaning, query an appropriate site on the Internet, then transform the weather report into audio so that they can "speak" the answer. Pretty impressive if you ask me.

MIDI Devices

Although it's often difficult for listeners to discern, humans do not directly create much of the music we listen to today. Many sounds, such as sweeping violins or pulsating drums, emanate from synthesizers that use software to generate musical tones electronically.

To organize electronic sounds into a song, musicians and composers need a way to direct a synthesizer to play a given note (e.g., middle C), with a given sound (e.g., saxophone), at a given time (e.g., first beat of a bar), for a given duration (e.g., a quarter note). That's where Musical Instrument Digital Interface (MIDI) comes into play. MIDI is a communication protocol that integrates MIDI-compliant instruments with specialized software packages to record and play music (see Figure 6.12).

---

[11] Beware: We rely solely on the good graces and professionalism of the vendors involved to ensure that these devices don't "eavesdrop" inappropriately.

FIGURE 6.12   MIDI music.

ADVANCED TOPIC: COMPUTER VISION

One of the goals of the IT industry is to make computers as simple to use as possible. One way to achieve such an objective is to develop interfaces that are intuitive and natural. We've already seen some examples of this: smart devices respond to verbal commands and deliver information as spoken words.

Another way to mimic human behavior is by simulating vision. Specifically, technologies that allow computers to "see" the world as humans do. We've already mentioned one example of this capability: software that can unlock smartphones using facial recognition technologies.

There are, however, many other uses of this functionality: filtering software could identify inappropriate images (e.g., violent or pornographic) before young viewers might see them. Robots could locate a specific package (among many) and place it in the correct storage bin inside a warehouse. And quite possibly, the most futuristic use of this technology is in self-driving cars.[12]

But simulating vision is not as easy as it appears. For example, consider the street scene depicted in Figure 6.13.

As humans, we immediately recognize objects like people, cars, traffic signals, road signs, etc. Thus, while driving, we would (hopefully) take appropriate action to avoid hazards as we steer ourselves to our destinations.

However, as noted during our discussion on digitization in Chapter 3, images processed by computers appear as a series of pixels or dots. This begs the question: How does a self-driving car interpret the world, given that it only receives images comprised of pixels?

---

[12] There are many other commercial and governmental uses of computer vision, some of which may be an invasion of our privacy. Consider that using street cameras in conjunction with facial recognition software, police departments can know that yesterday evening you walked to your local deli to buy milk. Scary.

FIGURE 6.13 A typical street scene.

(Editorial credit: Sean Pavone/Shutterstock.com.)

In practical terms, a self-driving car's computer must:

- Acquire images from one or more onboard cameras
- Determine the types of objects in the image using a variety of techniques, including Artificial Intelligence (AI)
- Determine which objects might serve as directives (e.g., stop signs, traffic lights)
- By processing successive images (along with input from other sensors such as radar),[13] determine which objects are moving and compute their trajectories (i.e., is the car going to collide with a bike that might be traveling across the street)
- Process all the above input and respond accordingly.

Please keep in mind that the processing described above must be accurate and occur in real-time (i.e., as fast as possible). As of this writing, self-driving cars are on the horizon, and they will revolutionize the transportation industry and the way we view automobile ownership.

One last comment about self-driving vehicles: there is a moral issue that, as a society, we'll need to address. Consider a situation wherein a self-driving car's onboard computer

---

[13] Self-driving cars integrate many technologies such as GPS, radar, and lasers.

determines that a collision with multiple vehicles is unavoidable. Should the software direct the car to take action that will minimize the risk of injury to its occupants? Or should its programming be more altruistic and try to reduce the overall risk to everyone involved?

Stated another way, should a self-driving car's software take action that might increase the potential of injury to its own passengers to reduce the damage to other vehicles? Should lawmakers stipulate that decision by statute? Or should this choice be a configurable option set by the owners of such vehicles? Food for thought.

SUMMARY

This chapter reviewed several classes of peripheral hardware, including input, output, and combined input-output devices. We discussed common components such as keyboards and mice and more leading-edge technologies such as 3D printers and computer vision.

In the next chapter, we'll focus on networking.

CHAPTER 7

Networking

The Art of Communication

Getting information off the Internet is like taking a drink from a fire hydrant.

MITCHELL KAPOR

INTRODUCTION

Networking is an overloaded term that can mean many things, such as expanding our list of professional contacts, mingling at a business conference, or ensuring that all our current doctors are members of a new medical insurance policy. However, in the world of IT, this term has a specific definition:

> *Networking* is the ability to connect two (or more) electronic devices with the express intent to share data or computing resources.[1]

In a previous chapter, we observed that computers that don't connect to anything are of little practical value. That said, there's no reason to have your computer connect to any other device if you don't want something from it. Or, some other system might require resources residing on your computer, and you're willing to grant it access.

Those two scenarios represent the most common relationship among networked components. We refer to it as the *client-server* model. The requesting computer is the *client*; the device supplying the resource is the *server*.[2]

---

[1] We will defer discussing security concerns until later in the book.
[2] Note that among communicating devices these roles can reverse as needed.

The *client-server* model is a communications framework that pairs a service requestor with a service provider via a network connection. The requestor is the *client*; the provider is the *server*.

Note that the relationship among the servers is logical and forms at the point of request. For example, when computer A requests data from computer B, A is the client; B is the server. However, computer B might also require data from computer A. In that case, the roles reverse, and computer B acts as the client and computer A as the server.

Believe it or not, you're already familiar with this model. For example, you might type HTTPS://WWW.WEATHER.GOV into your browser's address bar when you want the weekend weather forecast. At that point, your browser becomes the client by requesting the report; the National Weather Service's computer provides the data and assumes the role of server.

In the sections that follow, we'll peel back more layers of the networking onion so you can understand how this works in more detail.

NETWORK ATTRIBUTES

Before we dive too deep, we need to understand some terminology. To begin, we typically classify networks based on the following attributes:

Scale   Scale is a geographic characteristic. Networks can interconnect computers located within a single room or across the globe. (Think Internet.)

Size   The number of interconnected devices determines the *size* of a network, from minimally two to millions. (Again, think Internet.) We refer to each connected component as a *node*.

Accessibility   Accessibility refers to the "openness" of a network. Consider that, unless you're authorized, it's not likely that you'll gain access to top-secret government databases. Alternatively, many stores, businesses, and municipalities provide "open" networks for customer convenience.

Objective   This attribute refers to the type of services a given network might provide. For example, we are all familiar with general-purpose networks like the Internet. However, other networks deliver specific, focused services. We will see examples of these shortly.

The sections that follow define these attributes in more detail.

Network Scale

As per above, the attribute, *scale*, classifies a network by the geographic area it serves. The sections below discuss the various categories in order of scope. Please note that there is some overlap in the definitions.

PANs

A PAN, or Personal Area Network, defines the smallest geographic area. It refers to inter-connected personal devices. For example, do you own a smartwatch that connects to your smartphone? Does your fitness tracker transmit data to your laptop?

All devices on a PAN interconnect wirelessly. (It wouldn't be very convenient to have your smartwatch tethered to your laptop by a network cable.) Although not required, PANs may also interconnect with Local Area Networks (LANs) and Wide Area Networks (WANs) (see below).

LANs

LANs serve a limited geographic range such as a home, an office suite, a single building, or an apartment complex. LANs support both wired and wireless connections and vary in size from two interconnected devices to thousands, allowing nodes to share comput-ing resources and exchange messages. However, most LANs also provide a *gateway* that enables access to other networks.

Figure 7.1 depicts an example of a typical home LAN.[3] The router drives the network and allows devices to interconnect wirelessly or via a cable. All nodes may communicate with each other and share resources. For example, every device can route requests to the printer. In addition, all nodes can gain access to the Internet via the router.[4]

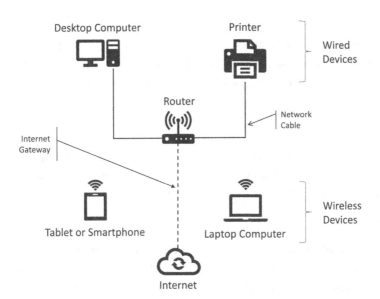

FIGURE 7.1   A typical home LAN.

---

[3] You may see this referenced as a Home Area Network (HAN).
[4] Discussed later in this chapter.

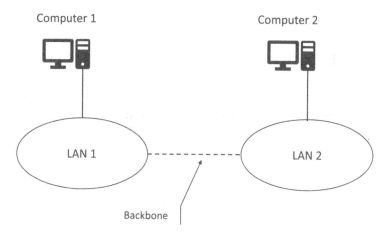

FIGURE 7.2   An example of a CAN.

CANs

A CAN, or Campus Area Network,[5] comprises two or more LANs connected by a *back-bone*. (A backbone is a network connection that joins multiple network segments.) A CAN is typically larger than a LAN but smaller than a MAN[6] (see below).

As depicted in Figure 7.2, the backbone allows Computer 1 to communicate with Computer 2 even though they reside on separate LANs. As a security consideration, administrators can configure the network to limit access to the backbone.

MANs

The term Metropolitan Area Network, or MAN, refers to a network that spans a large geographic area as large as an entire city. MANs can interconnect LANs and CANs to form a single cohesive network. The main distinction between a CAN and a MAN is the size of the area it serves.

WANs

Like a MAN, a WAN interconnects LANs into a single network. The difference is the size and diversity of the technologies involved. The most prominent example of a WAN is the Internet. (We'll discuss the Internet in detail later in the chapter.)

Network Objectives

Another essential attribute of a network is its *objective*. Most networks are general-purpose in nature in that their goal is to interconnect nodes so that they can perform day-to-day tasks, such as exchanging emails, copying files, and sharing resources (e.g., printers and plotters).

However, as described below, some networks are more narrowly focused and provide specific services.

---

[5] You might see this defined as a *Corporate Area Network*.
[6] IT professionals do love their acronyms.

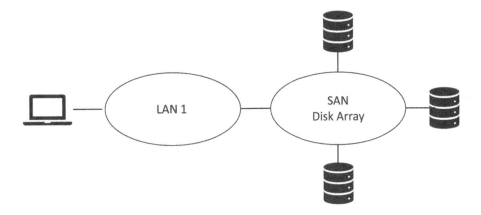

FIGURE 7.3   SAN storage array.

SANs

A Storage Area Network (SAN) provides access to bulk storage devices. (See Figure 7.3 for an example.)

A SAN makes a remote disk appear as internal storage within a computer. Thus, as depicted in Figure 7.3, the laptop interoperates with SAN disks as if they were locally installed devices. SANs allow organizations to centralize the management of external storage. Moreover, because it's network-based, SAN storage is available to any system anywhere globally (assuming viable network connectivity and adequate security).

EPNs

EPNs, or Enterprise Private Networks,[7] became viable in the 1970s using telecommunications technology. Like a CAN, MAN, or WAN, EPNs interconnect multiple LANs to form a single cohesive private network serving one group or organization's needs. Most EPNs rely on point-to-point leased communication channels[8] to interconnect geographically disparate sites. Although secure and reliable, they are costly to maintain.

VPNs

Since the advent of the Internet, EPNs have gradually given way to Virtual Private Networks (VPNs), which allow organizations to establish and maintain private networks whose nodes interconnect via the Internet. Thus, in place of expensive leased lines, sites interconnect securely and inexpensively using public networks.

To implement this service, VPNs use *encryption* and *virtual tunneling* protocols. Encryption is the ability to encode and transfer messages (data) so that only the sender and the intended receiver can interpret them. Virtual tunneling is the process of taking the sender's data packets (messages) and wrapping them inside the data packets of another service to mask them.

---

[7] Most IT professionals refer to EPNs simply as *private networks*.
[8] A typical example is an ISDN (Integrated Services Digital Network) connection.

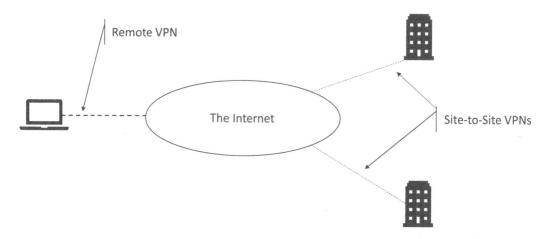

FIGURE 7.4   A VPN at work.

As depicted in Figure 7.4, VPNs rely on two types of connections: Remote Access and Site-to-Site. When users log into the VPN from a remote site, VPN software running on their PCs establishes a secure connection to one host. Despite traveling via public links, encryption and virtual tunneling ensure that the traffic remains private.

Site-to-site communication functions in much the same way. VPN software running at both sites ensures the privacy of the participants.

As an aside, are you tired of being tracked online? I know I am. To circumvent that, we can use VPNs to ensure privacy when "Surfing the Net." Third-party providers offer VPN services that allow you to connect to their VPN servers (using their proprietary software). Once connected, you become anonymous on the Internet because all your requests appear to originate from the vendor's VPN server rather than your PC.

Although they can offer reliable, anonymous communication, there is one caveat to keep in mind when using third-party VPN services. Before you license their software, you should determine whether the VPN provider maintains logs of your activity. The more data they retain about you, the less anonymous you remain. Although your search activity might not be visible in real-time, anyone with access to the logs can determine the sites you've visited.

We'll return to this topic when we discuss personal security in Chapter 13.

Network Access

The remaining attribute, *network access*, deals with system security. In general, to gain access to any electronic resource, the provider needs to determine your identity and based on that knowledge, decide what you can "see" and "do." Specifically, network access focuses on *authentication* (are you who you assert you are?) and *authorization* (what are you allowed to do?).

Authentication is the process of determining your identity. Using pre-established credentials such as login name and password—and other optional safeguards such as biometrics or codes[9] sent to your phone or email—security software verifies the identity of every user attempting to gain access to a network or system.

Authorization is the process of establishing the limits on what system resources individual users may access. For example, we would expect that hospital administrators are entitled to review your billing and payment history. However, they shouldn't have access to your medical records.

Network and system administrators don't establish and enforce security policies just to annoy you. They undertake these safeguards to prevent unauthorized use and modification of your data. Frankly, most people don't want to think about it, but data and identity theft are rampant. Unlike most people, I am always grateful when my banks and medical offices take prudent steps to thwart unauthorized access—despite the annoyance factor.

NETWORK ORGANIZATION AND INTERNETWORKING

In conjunction with scale and size considerations, administrators organize the structure of networks to maintain control over system resources. Specifically, they want to regulate how their networks interconnect with other networks—if at all.

Except in extreme cases (think CIA), this does not mean "locking down" every system resource. On the contrary, many organizations allow external access to trusted partners. For example, a warehouse company might permit suppliers to update inventory records after completing a delivery.

The sections below examine the most common examples of such network structures. However, as you read this material, please keep in mind that some of the designs discussed below are logical and may "overlay" physical networks.

Intranet

Like an EPN, an *intranet* is one or more interconnected networks controlled by one organization for the benefit of its users. For example, LANs, CANs, MANs, and WANs individually or collectively can function as an intranet. The key attribute is that the intranet remains private, available only to the managing organization's direct user base.

Extranets

Like intranets, *extranets* remain under the control of a single organization. However, they do allow limited access to external entities and users. Thus, organizations can selectively grant network privileges to whomever they want. For security reasons, network engineers only permit access to the extranet through single entry points called *portals*.

The Internet

Few individuals on this planet are unaware of the Internet. It has revolutionized many aspects of our lives: the way we work, play, shop, communicate, the list is endless. Its

---

[9] Called One-Time PINs (OTPs), the system generates these codes randomly and are valid for a short time and only one login attempt.

importance cannot be understated. Thus, to give it its full due, we will discuss it at length later in this chapter.

Darknets

Although accessible via the Internet, a *darknet* is a private[10] *layered* network that restricts accessibility to select groups or individuals.[11] You typically require specialized software and a personal invitation to gain access to these networks.

Despite their nefarious reputation, darknets support many constructive uses. For example, in countries that impose censorship, darknets allow their citizens anonymous access to the Internet.[12]

Sadly, as with any human endeavor, there are always individuals who can find ways to exploit anything. As a result, darknet technology does support pernicious activities such as black markets, human trafficking, and piracy.

NETWORK TOPOLOGY

Thus far, we've discussed many networking attributes such as size, scale, objective, and organization. However, we haven't described how system designers arrange the nodes (connected devices) on the network. We call this organizational structure the network *topology*.

> *Topology* characterizes the organization and hierarchy of the nodes residing on a network.

As we'll see shortly, there are several common topologies, each with different cost, performance, and reliability characteristics. As in all such cases, one is not better than another. Instead, each option satisfies a specific need in a spectrum of solutions.

Bus Network

Let's begin with a topology that we've already discussed: A Bus Network. Like an internal bus, all nodes connected to an external bus can communicate independently with any other connected node.

As depicted in Figure 7.5, a USB bus is a typical example of this design.

Star Network

With a design that looks like a bicycle wheel, all Star Network components connect to a central node called a *hub* (see Figure 7.6). All messages traverse through the hub before arriving at their destination.

---

[10] Darknets remain anonymous because they are not indexed by search engines (e.g., Google, DuckDuckGo, Bing, etc.) and are thus difficult—if not impossible—to locate.

[11] We will discuss overlay networks later in the chapter. However, you may recall that a VPN is an example of a layered network.

[12] One such service, the Tor Network, provides a browser that prevents tracking and masks its user's location.

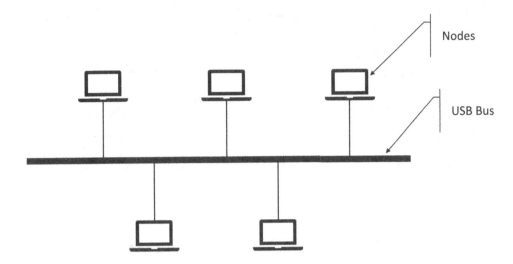

FIGURE 7.5   Bus network topology.

This is a common topology you've likely used without being aware of it. The Wi-Fi network in your home forms this topology because your wireless devices connect and exchange messages through a central hub (i.e., your router).[13]

Ring Topology

In a *ring topology*, every node connects to two nodes in sequence, thus forming a ring. Messages traverse the network in only one direction (either clockwise or counterclockwise) or bidirectionally if the implementation supports it. Each node in the network receives every message and will either process it or forward it to the next node.

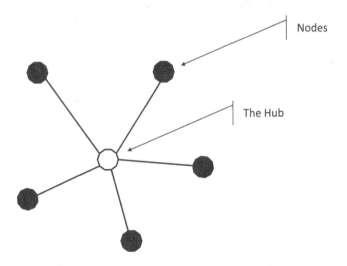

FIGURE 7.6   A star network topology.

---

[13] In network parlance, engineers often refer to the router as an *Access Point*.

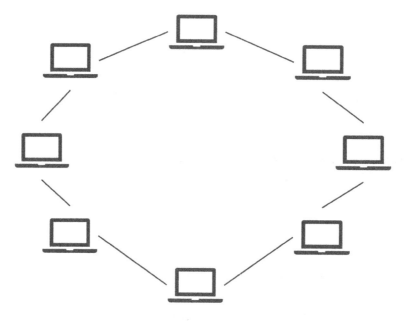

FIGURE 7.7   A ring network.

Figure 7.7 provides an example.

One advantage of a ring topology is that it outperforms a bus design during periods of peak load. One disadvantage, however, is that if one node fails, it might compromise the entire network.

Mesh Network

There are two types of *mesh networks*: *partially connected* and *fully connected*. In a *partially connected* mesh network, every node is aware of every other node on the network and tries to establish a direct connection with as many of them as possible. However, it's not always feasible for each node to connect to every other node in practice.

See Figure 7.8 for an example.

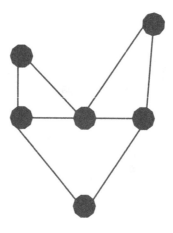

FIGURE 7.8   A partially connected mesh topology.

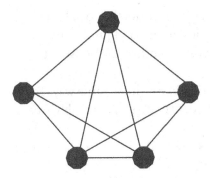

FIGURE 7.9   Fully connected mesh topology.

Partially connected mesh networks provide improved performance and reliability compared with star topologies. Consider that if one connection fails, a message can still reach its destination by traversing an alternate path.

In a *fully connected* mesh network (see Figure 7.9), every node connects to every other node.

Fully connected mesh networks improve performance because connected components can exchange messages directly without involving any intermediate nodes. Moreover, they improve reliability because the failure of one node does not compromise the entire network.

Hybrid Networks

There are many other network topologies, including *point-to-point*, *daisy chain*, and *hierarchical* (i.e., tree), that are beyond the scope of this book. In addition, network engineers can combine topologies to create *hybrid networks*. Figure 7.10 depicts an example of one such design that combines ring and star topologies.

NETWORK COMPONENTS

As endpoints, computers are the most critical nodes in a network: they are the devices that ultimately process messages. However, they alone do not create the network itself. As we will see, many other types of devices participate in a fully functional communication system to help transmit messages from senders to receivers.

To clarify this point, we can use a familiar example of a network: the US Postal Service. Consider that any sender can mail a package to any recipient who has a valid address. In this example, every mailbox (or P.O. box) is an endpoint (i.e., a computer). However,

FIGURE 7.10   A hybrid network.

mailboxes don't "move" packages. Instead, the Postal Service's infrastructure—comprising letter carriers, post offices, substations, and processing plants—collects, sorts, and transports parcels from one mailbox to another. Collectively, these resources form the backbone of the postal network.

Computer networks function in much the same way. The endpoints (computers) rely on a network's infrastructure to deliver messages reliably from an originator to a designated recipient.

In the sections that follow, we'll discuss the components that form the foundation of modern networks. Please keep in mind that although each device described below serves a specific purpose in managing network traffic, many of them are computers in their own right. Nonetheless, because of their focused functionality,[14] we tend to think of them solely as networking components.

Network Interface Controllers

Let's begin our discussions with your computer. PCs require a Network Interface Controller (or NIC) to connect to a network. Historically, NICs were available as accessory boards inserted into expansion slots inside the computer, thereby connecting to the system's internal bus. However, in modern PCs, the NIC often resides directly on the motherboard.

Regardless of its form factor, a NIC connects to external networks using cables (wired) or via radio waves (wirelessly). (Most modern PCs ship with NICs that support both.)

To support wired networks, NICs provide a *port* that accepts a cable connection. Most often, this is an Ethernet cable whose other end connects to a router. To enable wireless communication, the NIC contains circuitry that allows it to connect to radio-based networks such as Wi-Fi or Bluetooth.

One of the essential features of a NIC is that it ensures every network node has an individual address. Like a telephone number, every NIC contains what's called a *MAC Address* (Media Access Control Address) that uniquely identifies it, ensuring that senders can direct messages to specific recipients.[15]

As depicted in Figure 7.11, a MAC address is 48 bits in size and comprises two distinct parts: a 24-bit segment called an OUI (Organizational Unique Identifier) and a 24-bit

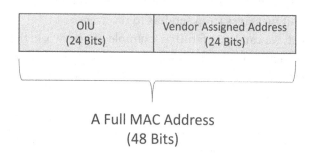

A Full MAC Address
(48 Bits)

FIGURE 7.11 MAC address components.

---

[14] We often refer to devices of this type as *appliances*.

[15] Networks also support the notion of *broadcast messages* that every node receives.

Vendor Assigned Address. Administered and assigned by an organization called the IEEE,[16] the OUI uniquely identifies every hardware vendor.

Vendors generate the second part of the MAC address and must ensure that the value they assign is unique for every instance of every product they manufacture. Thus, the combination of OUIs and vendor-generated addresses guarantees that every NIC is uniquely addressable.

Routers

The primary function of a *router* is to forward data packets between two or more networks.

Most of us are familiar with the routers that we use in our home networks. Typically supplied by your Internet Service Provider (or ISP), routers establish our home Wi-Fi networks and allow our devices to connect to the Internet.[17] See Figure 7.12 for an example. (As we'll see, routers serve as the backbone of the Internet.)

Hubs

The function of a *hub* is to form a private network, connecting two or more nodes without allowing access to other networks (or the Internet). When a hub receives a data packet from one of its connected nodes, it broadcasts that message to all other devices on its network.

Switches

A *switch* is like a hub in that it forms a private network among its connected nodes. It differs, however, in the way it routes traffic. Instead of broadcasting messages, a switch directs packets to a specific destination based on the recipient's MAC address.

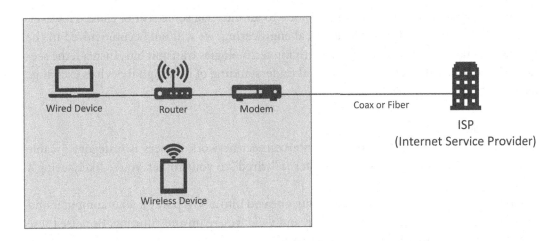

FIGURE 7.12   Typical home network.

---

[16] Institute of Electrical and Electronics Engineers.
[17] Not to get too technical, but a more accurate term for a home router is *residential gateway*.

Bridges

Bridges interconnect networks, forming a larger, composite network. Nodes on a bridged network can communicate with each other as seamlessly as if they resided on one network.

Modems

A *modem*, or MOdulator-DEModulator, allows network traffic to flow over transmission media not originally intended to support digital traffic (e.g., telephone lines). Although we still use the name (e.g., "cable modem"), modern broadband networks don't require them. Devices such as routers and bridges usually provide Internet access.

Firewalls

Security is always at the forefront of modern network design. To that end, network designers employ *firewalls* to prevent unauthorized access to networks.

Typically, firewalls are the entry ("touchdown") point into a network, and they implement the rules that grant or deny access. In most networks, Firewalls block access requests from all unknown devices.

NETWORK TECHNOLOGIES

As mentioned previously, we can construct networks with both wired and wireless technologies. In the sections that follow, we will describe the most common examples of each.

Physical Transmission Media

In wired networks, bits travel from one node to another via some sort of cable. However, because this is not a textbook on electrical engineering, we will not become mired in the technical details of how the various underlying technologies transmit bits. Instead, the sections that follow will provide a conceptual understanding of how digital devices exchange data using physical connections.

CAT 5 Cable

One of the most used technologies to interconnect network devices is Category 5 cable (called "Cat 5" for short). If your computer is "wired" to your router, you're likely using a CAT 5 cable.

CAT 5 systems are simple to install. Plug one end into a NIC port on your computer and the other into a LAN port on your router, and *voilà*, you're up and running. Because these connections are "hardwired" (i.e., physical), you typically don't require a password to gain access to the network.

Figure 7.13 provides an example of a CAT 5 cable.

Coaxial Cable

Coaxial cable (or "coax" for short—pronounced "coh-ax") is an electrical cable that contains an inner conductor (through which the bits move) surrounded by a shield (to minimize

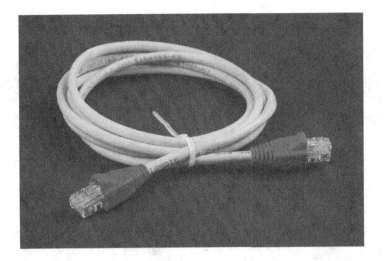

FIGURE 7.13   CAT 5 cable.

external interference). It's typically used in telephone networks, broadband connections, and television signal transmission. At home, you've likely used coax cables to connect TVs and other video devices to your cable or satellite network.

See Figure 7.14 for an example.

Fiber-Optic Cable
When using CAT 5 and coax, bits traverse the cable via electrical signals. With a Fiber-optic connection, bits move as pulses of light. Although there are performance advantages when using fiber, most users will opt for the convenience of wireless technologies (see below).

Figure 7.15 presents an image of a fiber-optic cable.

FIGURE 7.14   Coax cable.

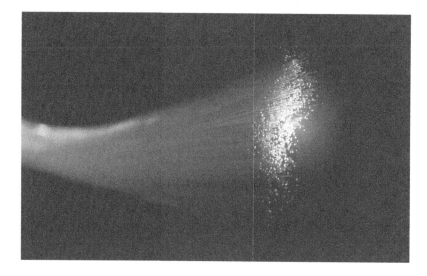

FIGURE 7.15 Fiber-Optic cable.

Ethernet

The prior sections discussed some technologies that *move* bits from one point to another. However, sending devices cannot just transmit bits at random and expect receiving devices to understand their meaning. The ability to successfully exchange messages requires structure and organization.

That's where *Ethernet* enters the picture. Ethernet is a *network protocol* that packages, manages, and controls data transmission (bits) over LAN connections.

Let's see how this works.

As depicted in Figure 7.16, the Ethernet protocol groups data into *packets*;[18] packets comprise several subsections: *header*, *payload*, and *trailer*.

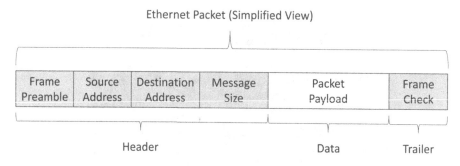

FIGURE 7.16 Ethernet packet—Simplified view.

---

[18] Technically, these are called Ethernet *frames*.

Like a mailing address, the header contains information that allows senders and receivers to exchange and track messages. It includes the following fields (among others):

Preamble	The preamble is a series of synchronization bits that senders transmit to alert receivers that a message is on the way. It ends with a specific bit pattern indicating that "the rest of the header is about to follow."
Source	This is the source (sender's) MAC address
Destination	This is the destination (recipient's) MAC address
Size	The length of the message in bytes. The Ethernet specification defines a minimum and maximum packet size.

Immediately following the header is the *payload*. This field contains the data that the sender wants to transmit to the receiver. For example, when you send an email, the payload field holds the message content.

The *Frame Check Field*, a component of the message *trailer*, is part of a process to confirm transmission accuracy. During packet preparation, the sender performs a mathematical computation called a Cyclic Redundancy Check (or CRC) and stores the value in the frame check field.[19]

Upon receipt of a packet, the receiver performs the same calculation and compares the result with the CRC included in the *Frame Check Field*. If the two values are not identical, it indicates that an error occurred during transmission. In such cases, the receiver may request the sender to retransmit the packet.[20]

Wireless Networks

Instead of using cables, digital devices can communicate wirelessly using radio waves.

Many of these technologies will be familiar to readers of this book (at least by name). We'll discuss the two most common in the sections that follow: Wi-Fi and Bluetooth.

Wi-Fi

Based on a set of standards,[21] *Wi-Fi* is a suite of wireless protocols that allow digital devices to connect to a *Wireless Access Point*.[22] A group of such connected devices forms a WLAN or Wi-Fi Local Area Network.

Wi-Fi specifications designate several radio frequencies that wireless networks can use. However, as of this writing, most implementations support either the 2.4 GHz or the 5 GHz bands (or both); each choice has both benefits and drawbacks.

For example, although cheaper, the 2.4 GHz band often suffers interference from common appliances (e.g., microwave ovens) that use the same frequency. In contrast, the 5 GHz

---

[19] A discussion of the actual CRC algorithm is beyond the scope of this text.
[20] There are other error detection and correction techniques used in data communication.
[21] Wi-Fi technology is based on the IEEE 802.11 family of specifications.
[22] For home networks, routers serve as the wireless access point.

suffers less interference but is usually more expensive, and its higher frequency reduces its range.[23]

One note. "Wi-Fi" is a trademark of a non-profit group called The Wi-Fi Alliance. This organization restricts the usage of the logo to certified products only. Thus, security issues aside, any Wi-Fi-compliant device can connect to any Wi-Fi-compliant *access point* anywhere in the world.

Bluetooth

Another prevalent wireless technology you've likely used is *Bluetooth*. For example, I'm sure most of you allow your smartphones to connect to your car's audio system so you can make hands-free telephone calls. That connection is Bluetooth-based.

Designed to enable mobile devices to connect to stationary components, Bluetooth technology supports a limited range (10 meters[24] or so); nonetheless, it's ideal for use in low-power devices like smartphones and headsets.

HOW DO NETWORKS WORK?

Okay, we've just covered a lot of ground, introducing numerous technologies, definitions, and specifications. However, we still haven't discussed how network nodes can accurately transmit messages from random senders to specific recipients. We call the "glue" that makes that happen the *Network Layer Model*.[25]

The next three sections will present how network models have evolved from conceptual designs into practical implementations.

The OSI Model

Let's begin by describing a general abstraction for network design called the *Open Systems Interconnection* (OSI) model. As depicted in Figure 7.17, the OSI model defines seven network layers (collectively called a *stack*). From bottom to top, they are the Physical Layer, Data Link Layer, Network Layer, Transport Layer, Session Layer, Presentation Layer, and Application Layer.

As the arrows indicate, each OSI layer on the sending side *logically* communicates with its corresponding layer on the receiving side. However, this is only true in the abstract. In reality, only the physical layer moves bits (data, messages) from the sender's device to the recipient's device.

Let's see how this works.

Under the OSI model, communication begins when an application (like a browser or an email program) wants to send a message. The *Application Layer* packages the message and hands it off to the next layer, in this case, the *Session Layer*.[26] The data packet progresses through all the intervening layers on the sending side until it arrives at the *Physical Layer*.

However, as each successive layer processes the message, it appends control information to the packet intended for its counterpart on the receiving side. You can think of this as

---

[23] Of late, price has become less of an issue. Most modern routers support both the 2.4 GH and 5 GH frequency bands.
[24] Or, if you must, about 32.8 feet in Imperial Units.
[25] Yes, another example of "layering."
[26] This is not entirely accurate. The Session Layer might not take part in every message transmission.

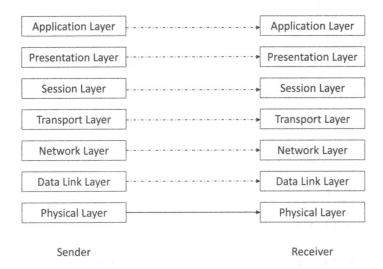

FIGURE 7.17   The OSI model.

envelopes within envelopes. That is, each layer accepts an "envelope" from a higher layer then places it into its own "envelope" before passing it on.

When a data packet is ready for transport, the Physical Layer on the sending side transmits the bits of the enclosing "envelope" to the Physical Layer at the destination (using one of the bit-moving technologies described above). Upon receipt, the receiver performs the complement processing: each layer removes its "envelope," processes the extracted message as required, and then hands it off to successive higher layers until it arrives at the intended application program.

What we've just described might seem like a lot of work to send a single message. On the contrary, as we'll see shortly, this approach ultimately reduces communication complexity. Moreover, we won't need to discuss every layer's responsibility because, in practice, we don't need to implement the full OSI model.

The following two sections discuss the most common network protocols.

The DARPA Network Model

In 1973, the US Defense Advanced Research Projects Agency (DARPA) began a research initiative to develop methods to simplify the interconnection of disparate networks. Out of that study, a streamlined network model emerged.[27]

As depicted in Figure 7.18, the DARPA model contains only four layers. Again, from bottom to top, they are Network, Internet, Transport, and Application. As in the OSI model, each sending layer logically communicates with its counterpart on the receiving side.

The role of each layer is as follows:

Network Layer   The Network Layer corresponds to the Physical and Data Link layers in the OSI model. This layer transmits and receives bits.

---

[27] Not to mention the genesis of the Internet.

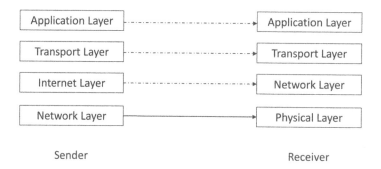

FIGURE 7.18 The DARPA network model.

Internet Layer This layer moves packets from a source to a destination. It's equivalent to the OSI Model's Network Layer.

Transport Layer This layer corresponds to its namesake in the OSI Model; it ensures reliable end-to-end delivery of data packets. (We'll understand how the Internet and Transport layers cooperate shortly.)

Application Layer The Application Layer corresponds to the top three layers (Session, Presentation, and Application) in the OSI model. It represents how applications logically communicate with each other.

Although simplified compared to the OSI stack, the DARPA model is still just that: a model. It still needs some practical streamlining. In the next section, we'll explore one of the most popular networking designs to have evolved from this paradigm.

The TCP/IP Stack

One of the most common networking models in use today is the TCP/IP stack. It's a suite of protocols and standards based on the DARPA paradigm that serves as the foundation of the Internet. Figure 7.19 shows the TCP/IP stack alongside the corresponding layers of the OSI and DARPA models.

Before we continue, one note on terminology: IP professionals often use the terms "Network Layer," "Data Link," and "Data Link Layer" interchangeably. Moving forward, we will only use the term "Data Link Layer" to eliminate any confusion.

For the discussion that follows, please refer to Figure 7.20.[28] It portrays the movement of an email message sent from one server to another.

Data Link Layer

As mentioned earlier, many individual devices comprise a network: routers, switches, hubs, etc. The Data Link Layer ensures that packets move reliably from one node to the next. Some of the Data Link Layer protocols highlighted in Figure 7.20 (e.g., Ethernet and Wi-Fi) should be familiar.

---

[28] Please note that this figure only presents a few examples of the protocols available for use at each layer.

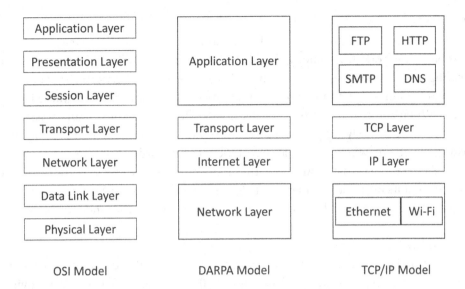

FIGURE 7.19   The TCP/IP "Stack."

The IP Layer

IP is an acronym that stands for Internet Protocol. The IP Layer moves data packets from one network to another.[29] (Hence the term "internet.") In other words, it manages all the *intermediate* hops a message undergoes until arriving at its destination. As depicted in Figure 7.20, that may include changes in the Data Link Layer protocols (e.g., Ethernet, Wi-Fi) as the message moves from node to node through all the intervening networks.

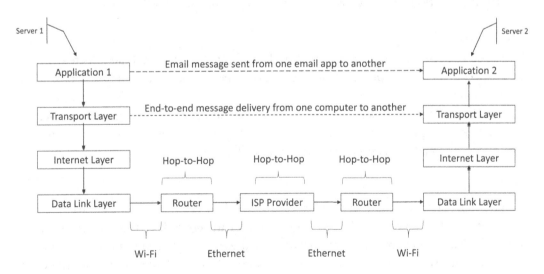

FIGURE 7.20   Example TCP/IP packet movement.

---

[29] Please recall the definition of a WAN from earlier in the chapter.

The TCP Layer

The TCP, or Transport Control Protocol, Layer manages message delivery from a source to a destination. In network terminology, the TCP layer establishes a *channel* between two communicating devices and ensures the accurate delivery of packets between the endpoints.

Let's take a moment to summarize the separation of responsibilities of these two critical layers. When transmitting data between nodes, the IP layer must manage issues such as changes in the underlying protocols, packet retransmissions when failures occur, and payload resequencing when packets arrive in the incorrect order.

The TCP layer ensures that regardless of the number of intermediate hops, the message ultimately arrives at its intended destination. It also maintains the channel between the communicating entities and manages flow control (e.g., a receiver can tell the sender to "slow down").

Application Layer

The Application Layer represents the software components that exchange messages via the other layers. For example, as depicted in Figure 7.20, APPLICATION1 (the sender) provides its local Transport Layer a chunk of data (the message). The TCP and IP layers append their routing information to the message and present it to the Data Link Layer. The Data Link Layer forward the message—via multiple intermediate hops using the information provided by the TCP and IP layers—to the Data Link Layer on the recipient's host. Upon its arrival, the message is "unwrapped" and delivered to APPLICATION2 (through the TCP and IP layers on the destination host).

The TCP/IP model makes it appear to both applications that they are communicating directly. More importantly, they can remain unaware of the myriad details associated with data transmission.

In the next section, we'll see how the Internet leverages the TCP/IP stack.

THE INTERNET

Unless you've been hiding in a cave, you're well aware that the Internet has affected almost every aspect of life on this planet. Moreover, a sizable percentage of this planet's population has access to it.[30]

As a practical matter, this means there are a large number of interconnected devices.[31] Thus, before we begin our discussion of this worldwide phenomenon, we need to understand how we can address every node on this global network.

What's DNS?

To understand how the Internet handles addressing, let's begin with an example: the cell phone network. Conceptually, when we want to speak with a friend, we think something like: *I want to call Joan.* But to reach her, the cell network uses her telephone number, not her name.

Although we rarely (if ever) concern ourselves with the underlying technology, there are many "moving parts" involved in initiating a cellphone call. For example, your friend can

---

[30] Almost 60% at the time of this writing.
[31] Some estimates predict that by the year 2025, there will be over 35 *billion* devices connected to the Internet.

change telephones, telephone numbers, and locations. So, how does the network track the information associated with every user to complete every call reliably?

First, let's consider what happens when your friend changes telephone numbers. In the old days, you'd typically dial directory assistance (411) and ask the operator for your friend's new number.[32] Today, you'd likely perform a search using the Internet. In either case, the result is that you associate your friend's name with a new telephone number. From a practical perspective, you would now dial a different number to call the same individual.

Next, what if your friend purchased a new smartphone but retained the same telephone number? In this case, the cellular provider must "map" the old telephone number to the new phone. In other words, you'd dial the same number but connect to a different device.

So how do these two examples relate to the Internet?

Well, every device connected to the Internet has a unique address called an *IP*[33] *Address*;[34] they are of the form NNN.NNN.NNN.NNN.[35] For example, the IP Address for NASA.GOV is 23.22.39.120.

The Internet's TCP/IP protocol requires that message packets specify destinations using IP Addresses. Nonetheless, I'm sure you've never typed an IP Address[36] into your browser; instead, you'd typically enter a domain name like NASA.GOV.

Returning to our telephone analogy, a domain name is like your friend's name, and the IP Address is like a telephone number. Thus, when you enter a domain name into an address bar, your browser "looks up" the name and maps it to a number (i.e., NASA.GOV to 23.22.39.120), and uses the number to connect to the other computer. (Again, you think "name;" the system uses the "telephone number.")

So how is this done? Must every browser store the IP Address of every domain on the Internet? That would be highly impractical.

Fortunately, the Internet provides a "411" service called Domain Name System (or DNS for short). When a domain "registers" on the Internet, DNS "maps" its name to its public IP Address. So, when you type a domain name like NASA.GOV, the browser's first task is to call out to a DNS server to acquire the correct IP Address of NASA's server.

Okay, so that covers the mapping of domain names to IP Addresses. But what about mapping IP Addresses to specific servers?

Please recall that computers use NIC cards to connect to networks and that every NIC card has a unique MAC address. For most networks, routers manage this mapping by maintaining internal tables that associate external IP Addresses with internal MAC addresses. Thus, when processing an inbound message, the router queries the destination

---

[32] I know most smartphones capture and save telephone numbers, so we don't dial "411" much anymore; just go with me on this one.

[33] Yes, this is the same "IP" as in Internet Protocol discussed earlier.

[34] Not to confuse the issue, but there are public and private IP addresses. As their name implies, public IP Addresses are known throughout the Internet. Alternatively, we use private IP Addresses to identify nodes on private LANs. For example, your home router will likely have a private IP Address of 192.168.1.1 that only has meaning on your network.

[35] This is Version 4 ("IPv4") of the IP Address format. Due to the explosion of devices connecting to the Internet, IPv4 is quickly running out of addresses. In response, the industry is rolling out Version 6 ("IPv6"), which can support 2^{128} addresses.

[36] To visit NASA's website, you could enter the IP Address into your browser instead of NASA.GOV.

IP Address (contained in the message's header—see above) and routes it to the appropriate server based on its associated MAC address.

Please note this solution also handles the case of server replacement. Referring to our ongoing example, this is the equivalent of your friend purchasing a new cellphone but keeping the old telephone number. (Remember, the new computer would ship with a new NIC card with a different MAC address.) In this case, the router would direct messages to the new server by associating the external IP Address with the new NIC card's MAC address in its mapping table.

One last comment on IP Addresses. Each time a device connects to a router on a home network, the ISP assigns it a *temporary* IP Address that can vary with each session. As a rule, such transitory addresses do not find their way into DNS registries. That's why it's difficult for a third-party server to locate your computer;[37] your temporary IP Address does not appear in the directory.

What's a URL?

As we have just seen, DNS helps us locate computers on the Web. Most of the time, however, we're not interested in *computers*. Instead, we want access to the *resources* they host.

For example, a typical YouTube server likely hosts thousands of media files. Unfortunately, knowing the IP Address of a YouTube server does not help us locate a specific video we want to watch. That's where Universal Resource Locators[38] (URLs[39]) come into play—they uniquely identify every resource hosted on a web server.

A URL comprises several distinct elements: a protocol, a hostname, and a resource identifier.[40] As an example, a generic URL might look something like this:

HTTP : / / WWW . SOMEHOSTNAME . COM / INDEX . HTML

The string, HTTP (Hypertext Transfer Protocol), is the protocol,[41] WWW.SOMEHOSTNAME.COM identifies the server,[42] and INDEX.HTML is the resource. (Please note that ":://" and "." serve as syntactical separators.)

For example, if we wanted to visit NASA's website, we'd type the URL HTTP://NASA.GOV into our browser's search bar. In this case, HTTP is the protocol, and NASA.GOV specifies the server. (In this example, the resource, a webpage, is implied.)

After we press the ENTER key, our browser would look up NASA.COM in a DNS server to determine its corresponding IP Address, establish a connection with NASA's server, and request the WWW (home page) resource. In response, NASA's server would send a series of commands instructing your browser how to display the requested information on your screen.

---

[37] But this is by no means impossible. That's why you should remain hypervigilant about network and computer security. More on this later in the text.

[38] Considered the inventor of the World Wide Web, Tim Berners-Lee defined URLs in 1994 when he issued RFC 1738. The world has never been the same.

[39] When speaking, we often refer to a URL as a "web address."

[40] More accurately, a URL is one type of Universal Resource Identifier (URI), which, when fully defined, contains five components.

[41] Formally, this is called the *Scheme*.

[42] Formally, this is called the *Authority*; it has three subcomponents.

As another example, consider this URL:

$$\text{HTTPS://WWW.YOUTUBE.COM/WATCH?v=Wfoy\_OvNDvw}$$

Hosted on YouTube, this URL will launch a video of a NASA spacewalk. Note the added component, "/WATCH?v=Wfoy_OvNDvw", appended after the hostname. This syntax identifies the exact video we want to play.

Though less commonly used, URLs can specify many other services and resources beyond webpages and videos. For example, you can download files (FTP), send emails (MAILTO), and log in to remote servers (TELNET), just to name a few. The key point is that because the structure of URLs is flexible, the industry can easily introduce new protocols and resources as the need arises.

NETWORKING EXAMPLE

Okay, it's time for a comprehensive example. Again, referring to Figure 7.20, let's trace the processing that occurs when you visit a web page.

You begin by launching your favorite browser (e.g., Firefox, Edge, Chrome, etc.). After typing a URL in the address bar, let's say, SOME-DOMAIN.COM, you press ENTER. Using a DNS server, the browser determines the destination computer's IP Address and passes the request to communication software on your local computer.[43]

Next, the TCP layer on your computer establishes a connection (channel) with the destination server. As part of this process, the intermediate network servers (hops) are identified. (Remember, your PC doesn't have a direct network connection to SOME-DOMAIN.COM. The TCP layer must find a path from your machine to it.)

At this point, the IP and Data Link layers on your computer transmit the message to the first hop. (This is the router in Figure 7.20.) Using the IP information included in the message header, the router forwards the message to the next node in the communication chain.

The hop-to-hop processing continues until the message arrives at the Data Link Layer residing in the destination server, at which point the network layers on the receiving computer "bubble up" the message to the webserver that will respond to the URL request.

Yes, I glossed over many of the tasks and details involved in message routing in this illustration. Nonetheless, it conveys the point: each software layer in the network stack has specific responsibilities. More importantly, applications (the browser and webserver in this example) don't have to become mired in the technical tedium associated with data transmission.

ADVANCED TOPIC: IoT

As most readers of this book are aware, the Internet has connected the world in ways we hadn't even thought of a few years ago. We can work, chat, compose music, and play games with folks around the world as if they were sitting in the same room. But that's only the beginning.

---

[43] We'll discuss how this happens in the chapter on Operating Systems.

You might be familiar with smart home technology—the ability to automate lights, thermostats, alarms, sprinkler systems, etc. But what about coffee pots, cars, refrigerators, toasters, alarm clocks, activity trackers, and, well, everything else.

The next level of interconnectivity is in its nascent stages. Called the Internet of Things (IoT), it promises to connect every device on our home networks to the rest of the world via the Internet. With IoT, every digital device will eventually be able to communicate with each other.

For example, would it be nice if, in addition to waking you up at the appropriate time, your alarm clock signaled your coffee maker to begin a brew cycle, started your shower so the water would be warm, and after querying your calendar, sent your car's navigation system the location of your first stop? Would you find it convenient if your refrigerator placed an electronic order with your grocery store? What about a smart traffic system that could reduce congestion by automatically rerouting cars via their navigation systems? Would it benefit cardiac patients if an IoT device could track their heart and respiratory rates so that their doctors could adjust their pacemakers in real-time?

Is this a good thing? It can be—if controlled and used appropriately.

Is there a downside? Yes. I know I'm harping, but so long as humans reside on this planet, one or more of them will determine how to exploit IoT's power for illegitimate purposes.

SUMMARY

In this chapter, we learned that there are a lot of moving parts in a computer network. And although most of us won't become network engineers, there are several key points to keep in mind:

- We classify networks by size (LAN, WAN, etc.)

- Network designs can serve specific purposes (SANs, VPNs, etc.)

- Topologies (Ring, Mesh, etc.) organize nodes on a network

- In addition to computers, many other devices connect to networks (smartphones, routers, gateways, etc.)

- Every network device has a NIC that ships with a unique MAC address

- Most devices have a public IP Address that maps to their private MAC address

- Web addresses (URLs) map to IP Addresses

- Computer networks comprise several conceptual layers (Data Link, Internet, Transport, and Application)

- TCP/IP is the most used network "stack"

- The Internet is the backbone of an interconnected set of independent networks

Given the above, I hope the next time you *google* "movies near me," you'll appreciate how many devices might be involved in delivering that information to you.

CHAPTER 8

Software

The Elusive Enigma

That's the thing about people who think they hate computers. What they really hate is lousy programmers.

LARRY NIVEN

INTRODUCTION

In previous chapters, we've discussed the more tangible components of computer systems. The stuff we can see, touch, plugin, and pound with our fists when things aren't working correctly.

However, that's not what jumps to mind for most folks when they give thought to computers. Most people think in terms of applications—Waze, Grubhub, MS Word, etc.—rather than CPU, bus, and device drivers. And rightly so. Like cars, we want computers to take us where we want to go without worrying about the annoying details.

We'll approach software like our hardware discussions. We'll use this chapter to paint a conceptual overview of this elusive enigma. In subsequent chapters, we'll "divide and conquer" as we dive more deeply into this vast, limitless ocean.

WHAT IS SOFTWARE?

As I alluded to above, software is the public face of computers. It's not an exaggeration to say that applications bring the world to our fingertips. From news to movies to chatting with friends or engaging in virtual doctor visits, it's (mostly) software[1] that brings all that to life.

But how?

---

[1] Yes, you need hardware; but, for the most part, it's software that "brings computers to life."

Let's begin with a definition.

Software is an organized arrangement of data and instructions that, when executed, direct the operations of the underlying hardware to accomplish a particular purpose.

This definition shouldn't be revelatory to you if you've read the preceding chapters. Nonetheless, we should take a moment to review it.

Let's begin with the first part. In Chapter 4, we learned that computer programs reside in memory and comprise two main sections: instructions and data. Instructions direct the hardware's operations and determine how to manipulate the data.

The second part of the definition, "to accomplish a specific purpose," is the reason why we have thousands of apps and not just one. Whether it's a spreadsheet, word processor, or movie player, each program serves a specific need. In other words, if you want to file your taxes, you use a program like TurboTax, not Photoshop.

Although accurate, this definition does not provide sufficient detail. So, let's peel back some layers of the onion in a way that won't bring tears to your eyes and see how we organize software.

SOFTWARE LAYERS

Broadly speaking, we can divide software into two major categories: *system* and *application*. For the most part, *system software* runs behind the scenes. Its responsibilities include managing the hardware, protecting the system from unauthorized access, and performing routine maintenance.

The term, *application software*, refers to programs designed for users that perform clearly defined functions. These are the software packages you use every day. Some examples include browsers, email apps, word processors, social media services, and eReaders (which you might be using right now to read this book).

However, these are broad classifications that require further refinement. Throughout this text, we've seen repeated examples demonstrating that the world of computing resides in *layers*. As depicted in Figure 8.1, software is no different.

Let's discuss the layers from bottom to top.

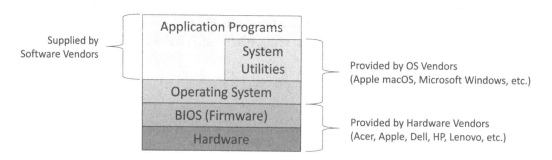

FIGURE 8.1   Software layers.

Firmware and BIOS

Previous chapters have already discussed hardware in detail; so, let's not spend any additional time on the subject. However, I'd like to take a moment to discuss the *BIOS* layer.

Supplied by hardware vendors, BIOS, an acronym that stands for *Basic Input/Output System*,[2] introduces us to a type of system software called *firmware*.

> *Firmware* is software integrated into the hardware that controls the operation of the underlying circuitry.

Conceptually, *firmware* is like system and application software. The difference is the level of control. In modern digital devices, engineers design circuitry so that software (i.e., firmware) controls its operation. Thus, designers can modify and extend firmware to meet changing demands without replacing or upgrading the underlying hardware. In some sense, this means that even hardware has become programmable.

As noted above, *BIOS* is a type of firmware. It's pre-installed software that controls system initialization,[3] identifies connected devices (e.g., mouse, keyboard, disk drives, etc.), and provides run-time support for the operating system (OS) (discussed below).

As with all firmware, BIOS usually resides in a non-volatile data store (we've already discussed one example of this: flash memory) and doesn't change all that frequently. (Nonetheless, during system upgrades on your PC, you might receive a notification that a "BIOS update" took place.)

We won't spend any more time discussing firmware; its focus is too narrow, and it's so integrated into the underlying circuitry as to become indistinguishable from the hardware (at least for our purposes). Nonetheless, please keep in mind that because it's software, firmware allows for easy extension and upgrade without requiring hardware component replacement.

Operating System and Utilities

Moving up, the next layer in our software stack is the Operating System (OS) and its associated utility programs. Let's begin with the OS proper.

> An *Operating System* (*OS*) oversees and controls access to all system resources (both hardware and software) and manages the execution of all application programs.

Most digital computing devices (PCs, smartphones, smartwatches, tablets, etc.) rely on OSs[4] for overall command and control. And although you're likely familiar with some of the more popular products—MS Windows, macOS, Linux, Android, iOS—you might not be aware of what they do because you rarely interact with them directly.

---

[2] You might have also heard this referred to as System BIOS or ROM BIOS.

[3] Often called a *system boot*, this is the "start-up" sequence that prepares your system for operation whenever you power it up or "restart" it.

[4] Some devices ship with a "small" operating system called an *executive*.

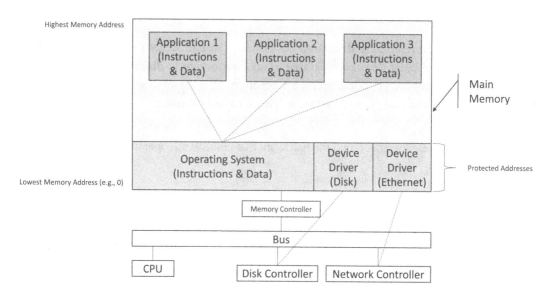

FIGURE 8.2   Operating system.

Working behind the scenes, OSs host applications and manage access to all resources. Consequently, we often refer to them as *platforms*.

Figure 8.2 depicts a representation of the internal state of your PC when it's running.

Let's focus on the box representing main memory. As part of system startup, the *boot process*[5] loads the OS into the lower part of memory. Enforced by the hardware,[6] this area is "off-limits" to all other applications; only the OS may access data residing in this section. Note that *device drivers*[7] load into this protected memory segment. In effect, they run as extensions to the OS.

From this protected location, the OS safeguards and manages access to all system resources. We'll see how this works in the Chapter 9.

As also depicted in Figure 8.2, all other programs execute in *application space*: any memory area not otherwise reserved for the OS. Each running program receives its own memory segment. It's the responsibility of the OS to ensure that applications don't exceed their allotted boundaries.

Most OSs also include numerous utility packages. That is, in addition to the system components that load into low-level memory during the *boot process*, OS vendors provide a suite of programs that help manage the system. Most of these utilities reside on disk, and you run them like any other application program.[8]

Utility programs come in many "flavors." For example, you may have used virus scanners, disk repair tools, system monitors that display CPU and memory usage, diagnostic programs that identify problems, and so on. Each OS vendor supplies a suite of tools that

---

[5] We will return to the boot process in Chapter 9.
[6] How this is accomplished is beyond the scope of this book.
[7] Discussed in Chapter 4.
[8] Indeed, they have the same structure as application programs. The difference is that they are part of the OS package and "understand" details of the OS that most general-purpose applications do not.

they believe serve their customers' needs. You can also purchase third-party utilities that replace, augment, or enhance the collection included with your system.[9]

Application Software

If the OS is the plate, applications are the meal. (This is the topmost layer depicted in Figure 8.1.) Ultimately, they are the reason that we purchase computers. For example, despite its advanced technology, how useful would an Ultra HD TV be if networks didn't broadcast any shows in that format?

The same holds true for computers. You could have the most advanced hardware and OS known to humankind, but it wouldn't be of much use if you couldn't run any applications on it.

Formally, we define an *application* as follows:

An *application* is a collection of one or more programs designed to serve a specific need.

There are more applications than we could cover in a volume of books. They range from:

General Purpose	Email utilities, word processors, music players, browsers, spreadsheets, social media apps, …
To the Specific	Banking apps, shopping apps, games, video streaming, billing applications, …
To the Even More Specific	Surgical software, autopilot systems, ground-penetrating radar software that categorizes subsurface structures, facial recognition, robotic artificial intelligence, …
To the Bespoke	Automated trading applications used by brokerage firms, custom order management systems, proprietary market analysis tools, …
To the Personal	I wrote a small program that allows me to package all the files (text, figures, logs, etc.) that comprise this manuscript to quickly ship them to the publisher as a single ZIP[10] file.

I could continue listing applications *ad nauseam*. The point is that if there is a need, it's likely that there's an application that can fill it. Moreover, despite its obvious triteness, the following statement is true: the types of software applications are limited only by human imagination.[11]

Please note that there are usually multiple product offerings in each category listed above (other than custom). For example, you might use the default email app that shipped

---

[9] *Caveat Emptor*: Make sure you know with whom you are dealing. Do NOT download anything unless it's from a trusted site/vendor and you're certain of its provenance and its *bona fides*. As we'll see in Chapter 13, malware is rampant on the Internet.

[10] That's another category of software: data compression.

[11] And cost, time, resources … unfortunately, software designers must deal with such practical considerations.

with your phone. However, if you took the time to search, you'd find scores of competing products, each with its own relative set of advantages and disadvantages.

ADVANCED TOPIC: MICROCODE

Earlier in the chapter, when we discussed the software stack, we indicated that the lowest layer was the BIOS (firmware). However, for most modern processors, that's not the complete story.

But for us to move forward with this discussion, we'll need to take a step back.

During the early years of digital chip design, electrical engineers "hardwired" the circuits. That means for every machine instruction, such as ADD or MOV, there was a predefined path in the circuitry. This approach was intentional; hardwired circuits are extremely fast. However, as always, there was a "downside."

As application programs became more elaborate, they placed additional burden on the underlying hardware. In response, the machine instructions designed into a typical CPU became correspondingly complex, which led to more intricate circuitry. This further complicated the design, manufacture, and testing of new generations of hardware.

So, how did the industry address this issue? Through *layering*, of course.

Engineers began designing electronics controlled by programming. This software layer, called *microcode*, allows designers to define and refine circuits without "rewiring" the chip.[12]

Microcode is a layer of software that executes the machine instructions in the hardware.

Microcode resides in customized, high-speed memory controlled by the CPU. When executing a machine instruction (e.g., ADD or MOV), the hardware maps the opcode to a sequence of microcode instructions and then runs them.

CPUs that employ microcode do suffer a small performance penalty because they require one additional level of translation. However, the gains in flexibility and extensibility more than justify this design choice. For example, engineers can correct design flaws by altering the microcode, and existing systems can receive new/revised hardware instructions via a simple software upgrade.

SUMMARY

In this chapter, we presented a conceptual view of computer software. We discussed how it comprises several layers and how the lower layers create a platform that serves as the foundation for application execution. In succeeding chapters, we will expand on this material.

Next up: A deep dive into the OS.

---

[12] By no means does this statement imply that advances in hardware technology no longer occur. On the contrary, the design of digital circuitry evolves continually.

CHAPTER 9

Software

Operating Systems

No one cares what operating system you run as long as it stays up.

BRUCE PERENS

INTRODUCTION

Although you may not fully understand what they do (yet), you are familiar with the most common operating systems (OSs): Google's Android, Apple's macOS, Microsoft's Windows, and Linux.[1] However, based on this chapter's introductory quote, you may be wondering why we're discussing OSs at all. I mean, if it's some behind-the-scenes component that "no one cares" about, why bother?

Well, to make the case, let's begin at the beginning. During the early days of digital computing, software developers had to program every detail: if they wanted a character to appear on a printer, they had to understand how the printer worked; if they wanted to write data to storage, they had to understand the underlying disk technology; if they wanted to add two numbers, they had to understand the machine language of the CPU.

This type of low-level, detail-oriented programming is tedious, time-consuming, and prone to error. (The more code developers write, the more errors they'll make. It's just human nature.) Obviously, it would simplify system development if we could package common processing (e.g., sending characters to printers) in ways that would allow developers to reuse it.

This type of thinking—packaging routine processing for reuse—is not new, nor is it the exclusive province of the IT world. For example, in the early days of the telephone system, human operators connected every call. It soon became apparent to telephony engineers that every person in the country would have to become an operator to satisfy

---

[1] These are the most common operating systems; there are scores of others used in the world of IT.

the projected call volumes at the then-current growth rate. Obviously, that wouldn't have been viable: if everyone served as an operator, who would be available to make the calls?

However, in effect, the industry developed a way to do just that: everyone became an operator. Specifically, using rotary dialers (initially) and then digital electronics, system engineers packaged "operator" processing such that everyone "connected" their own calls.

So, how might we accomplish this in OS software?

Well, we've already seen an example of this approach in Chapter 7. As we move down the network "stack," each successive layer performs tasks upon which higher layers can rely. Thus, developers working on the IP Layer don't have to rewrite the code required to transmit messages; they can assume that the Data Link layer will "move the bits" on their behalf.

We can extend this idea to operating systems: the OS is a software layer that provides "low-level" services to applications.

Another limitation of early computer systems was that they could execute only one process at a time. Thus, your program had to wait until mine finished. In contrast, most modern computer systems support the concurrent execution of multiple programs.[2] (We refer to this as *multitasking*; see the ADVANCED TOPIC section at the end of this chapter.) To implement such a feature, the system would need an "arbiter" to sequence process execution, prevent resource contention, and ensure that all applications remain "well-behaved." As we'll see, this is another feature OSs can manage for us.

However, before we go too far afield, let's refine our definition of an OS.

> An *Operating System (OS)* is a specialized program that oversees process execution and manages all the hardware and software resources available in a computer system.

You can think of an OS as a protective cocoon that insulates applications from the "harsh realities" of the hardware, offers many reusable services, and provides a safe and predictable execution environment for running processes.

We've already seen examples of how an OS manages hardware and software services, so let's take a moment to discuss what the phrase "predictable execution environment" means.

As I'm sure you're aware, hardware manufacturers continually expand their device offerings. It would become extremely annoying if applications began failing whenever you upgraded your system. That's part of the cocoon provided by an OS: it insulates applications from underlying hardware and system software changes. As depicted in Figure 9.1, a "complete solution" requires the integration and cooperation of both an application and an OS.

---

[2] You can see a list of all running tasks on your system by using ACTIVITY MONITOR under macOS, TASK MANAGER under Windows, and the ps command under Linux.

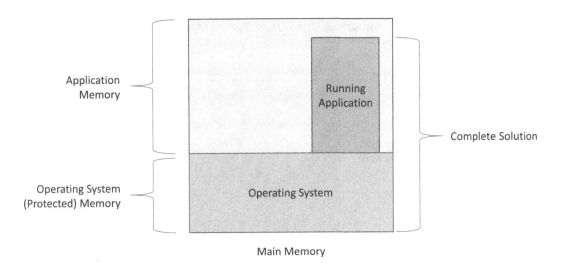

FIGURE 9.1   Combined Operating System and application program space.

WHAT DOES AN OPERATING SYSTEM DO?

As mentioned previously, OSs manage resources. But that statement is too broad. In the sections that follow, we'll discuss some of the specific responsibilities of a modern OS.

Process Lifecycle Management

One could argue that the most basic, essential task of an OS is process management. Whenever we launch an application, the OS must oversee its execution until it terminates.

However, before we proceed, let's note the distinction between a *program* and a *process*:

> A *program* is an ordered collection of computer instructions contained in a file. A *process* is the running image of a *program* as it executes in memory.

For example, when you CLICK on a *program*'s icon (e.g., a browser), the OS copies into memory the instructions contained in its disk file. The image now residing in memory is the *process*; it's only at this point that the program can "run."

Let's review how OSs manage the full execution cycle of a process.

Invocation

Initially, when you invoke a program, the OS must locate its file on disk and copy its instructions and data into memory. To track and manage the running process, the OS creates an entry for it in its *Process Table*.[3]

The *Process Table* contains information about all running processes that must remain readily available to the OS. It includes such items as the process ID (every executing process

---

[3] The name of this table may vary by operating system.

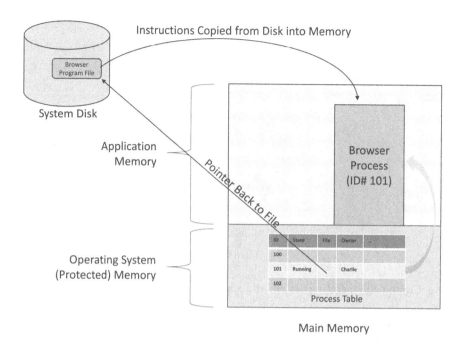

Main Memory

FIGURE 9.2   Process state after browser invocation.

receives a unique ID), the location of the program's disk file, the process's current state (e.g., *running*, *suspended*), and its priority (some processes may take precedence over others).

Figure 9.2 depicts the state of your system after you launch a browser but before it begins running. Note that the OS has loaded the browser's instructions and data into memory and has also entered relevant tracking information into the *Process Table*.

Execution

After completing the invocation tasks, the OS sets the process's state to "runnable," indicating that it may begin executing. At this point, the OS could allow your browser to start running. However, in modern computing environments, there's usually more than one process[4] that's ready to run at a given moment in time. Thus, the OS must now arbitrate among the competing demands of runnable processes with the limited resources available on the system.

This issue leads us to the next topic: *scheduling*.

Scheduling

The principal resource in any computer system is its CPU; it's the one resource required by *every* running process.

Scheduling is the method by which the OS grants CPU access to runnable processes.

---

[4] In addition to processes, there are other executable elements that the OS must manage (e.g., *threads*).

Given that a CPU can only execute *one* instruction at a time (as discussed earlier in the text), the OS may schedule only *one* process to run at any moment in time.[5] Once running, a process continues executing until it relinquishes the CPU (we'll see how this occurs below).

This processing model has several implications.

First, consider that the OS is a process itself; thus, whenever it's running—and therefore controls the CPU—all other processes must wait. When the OS completes its current task, it will schedule another process to run.

Next, this one-instruction-at-a-time limitation also impacts the OS itself. If an application process is running (such as process 101 in Figure 9.2), then all other processes *and* the OS must wait. (We'll see how the OS regains control of the CPU shortly.)

There are many types of OSs (see the ADVANCED SECTION below), and each employs different types of scheduling algorithms. Two of the most common are:

Priority Based   Every process has an assigned (or computed) *priority*. When scheduling, the OS selects the one that has the highest value.

Round-Robin Allocation   Each runnable process gets a "slice" of the CPU in turn.

In the next section, we'll see how the OS prepares a process for execution.

Context Switching

After determining which process should run next, the OS must suspend the currently running process and turn control over to the new one. For non-obvious reasons (we'll clear this up shortly), we refer to this task as *Context Switching*.

For example, as depicted in Figure 9.3, process 101 is currently running. When it's time for one of the other two processes to execute (let's say process 102), the OS will perform the *context switch* by saving the execution state of process 101 and restoring the execution state of process 102.

Please recall from Chapter 5 that a process has data and control information stored in system resources (e.g., registers) when it's executing. Before it can suspend a running process, the OS must preserve these values (collectively called the *Execution Context*) so that it can restore them when the process gets to run again.

> *Execution Context* is the state (value) of all system resources for a given process at a given moment in time.

For any process that's about to gain control of the CPU, the OS must restore its *Execution Context*. It does this by copying the process's previously saved values back into the appropriate system resource locations.[6]

---

[5] For now, let's assume that we're working with a system that contains only one CPU.

[6] There are other tasks involved in this process that are beyond the scope of this text.

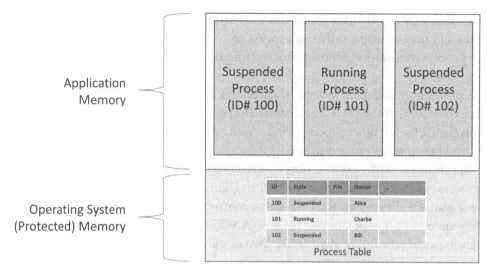

ID	State	File	Owner	..
100	Suspended		Alice	
101	Running		Charlie	
102	Suspended		Bill	

Process Table

Main Memory

FIGURE 9.3 Running and suspended processes.

This saving and restoring of the *Execution Context* is the derivation of the term *Context Switching*.

Context Switching is the saving and restoring of a process's *Execution Context*.

If you think that a *context switch* is an expensive operation (in terms of computing time), you're right. Overall, however, it minimizes system idle time and maximizes system efficiency. (If there's a process ready to run, it will gain access to the CPU.)

Process Suspension

Thus far, we've established two fundamental ideas in this chapter: the OS is the most important process in a computer system, and processes run until they relinquish control of the CPU. This begs the question: when the OS allows some other program to execute, how does it regain control of the system?

In broad terms, there are two ways by which running processes relinquish control of the CPU: through events called *traps* and *interrupts*.[7] However, we need to take a brief digression before we describe how these events fit into our scheduling model.

In terms of raw speed, there's nothing faster in a computer system than the CPU. Although it's difficult to perceive in human terms, peripheral devices such as disks, printers, and networks are significantly slower (by orders of magnitude) than the processor.

Thus, when a process requests service from a peripheral device, the OS has two choices. It can allow the requesting process to wait—and, by extension, the entire system because only one

---

[7] There is also the notion of *Exceptions*; we will not discuss these in detail.

process may execute at any given moment—until the peripheral device completes the operation. Or it can suspend the requesting process and allow another one to run while the peripheral device goes about its business. Most OSs adopt the latter approach to maximize overall system efficiency.

Let's discuss how the OS manages this processing.

As we have stated several times, the OS is responsible for managing the entire system. Consequently, processes cannot directly solicit services from peripheral devices. Instead, they must request the OS to intervene on their behalf.

For example, when a running process requires data residing in external storage, it cannot interact directly with either the disk or its controller. Instead, the process must issue a *read* request via the OS. We call this an OS *trap*.

When a *trap* occurs, the OS:

- Regains control of the CPU

- Manages the request or event (e.g., forwards it to a peripheral device)

- Sets the run state of the requesting process to *waiting*

- Determines which process should execute next

- Performs a *context switch* to allow the next scheduled process to run

OS traps occur for many reasons and can vary by system. The most common are input/output requests, error conditions, and exception handling (i.e., whenever a process "misbehaves," the OS must intervene).

The second way for the OS to regain control of the system is through an *interrupt*. Similar in concept to a *trap*, an *interrupt* is a signal[8] raised by a hardware device indicating that it needs attention from the OS.

For example, after a disk drive completes a read request, it will *interrupt* (signal) the OS; in response, the OS will:

- Regain control of the system;

- Determine which process's read request has completed and changed its execution state from *waiting* to *runnable*;

- Determine which process should execute; and

- Perform a *context switch* to allow the next scheduled process to run.

As we'll discover in the next section, the OS requires some help from the hardware to manage this processing.

---

[8] Please recall discussion regarding Bitwise Operations in Chapter 5.

Kernel Mode vs. User Mode

In the preceding sections, we noted that application processes cannot directly access system resources and must issue such requests via the OS. Clearly, the OS must have access to all system resources to service these requests. To implement this segregation of permissions, computer systems run in one of two states: *user mode* or *kernel mode*.

When the system is in *kernel mode*, the currently running process has unrestricted access to all system resources. Thus, because it's trusted, the OS is (usually) the only process that may run in *kernel mode*. On the other hand, application software is inherently untrusted, and therefore runs only in *user mode*, restricting its access to protected resources.

In modern computer design, the system enforces state changes from *kernel mode* to *user mode* preventing untrusted processes from accessing hardware resources and protected memory locations.

Process Termination

A running process can *terminate* (we also use the terms *stop, halt,* and *exit*) for many reasons. In the best case, programs exit when they complete their assigned task successfully. Additionally, users may stop applications at any time (for any reason).

However, there are occasions when the OS must intervene and terminate an application because it's "misbehaving." For example, a process may only retrieve data from memory locations within its assigned address space. If it tries to access an area that's "out of bounds," the OS will step in and terminate it.

The OS will also intervene whenever a process is about to execute an illegal instruction. A typical example of this type of (programming) error is when a process tries dividing by the value zero (0). In mathematics, this operation is "undefined" and is thus prohibited in the world of computers.

Regardless of the reason, the OS must "clean up" after a process when it terminates. Some examples of these "housekeeping" tasks include removing entries from the process table, freeing memory (so that other programs may use it), and aborting any in-progress operations (e.g., the OS will cancel previously issued disk read requests that have not completed).

The Idle Process

Computers are nothing if not logical. Consequently, divergence from the "expected" can often wreak havoc. So, what happens when there are no runnable processes in a system at some given moment in time? In other words, what should the OS do when there's nothing to do?

At system startup, one of the OS's first tasks is to launch what's called the *idle process*.[9] Its job is to do, well, nothing! However, it is always runnable (i.e., it's never *suspended*), it consumes no system resources, and it has the lowest system priority. Thus, when there's nothing else to execute, the OS runs the idle process.

---

[9] Most operating systems provide an *idle process*; however, its name can vary by product.

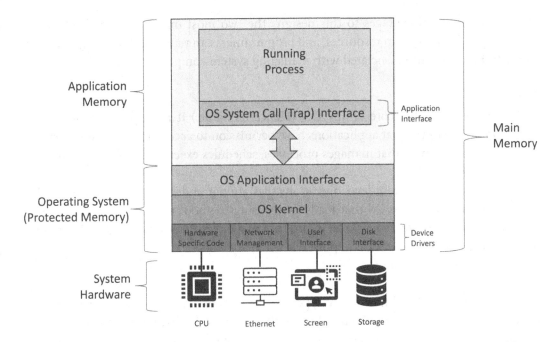

FIGURE 9.4 Logical architecture of an OS.

Resource and Hardware Management

Operating System Architecture

As we have observed, one of the most important tasks of an OS is to manage the underlying hardware. To understand how this process works, we need to discuss the structure of a typical OS; we refer to this design as its *architecture*.

Figure 9.4 depicts the logical architecture of a typical OS.[10]

As a reminder, the OS usually resides in an area of protected memory reserved for system software. This design prevents unauthorized access to the hardware and the protected memory segment itself.

The following sections discuss the architectural components in more detail.

Application Interface

As previously mentioned, when an application program requires a system service (e.g., reading data from a disk), it must call upon the OS to complete the task on its behalf. To initiate such a request, a process uses the *System Call* interface to issue a *trap* (see above) and "enters" the OS via the *Application Interface*. At that point, the OS *suspends* the application and begins processing the request (discussed in the next section).

The *System Call* interface is a collection of prepackaged functions that application programmers use to request specific services from the OS. Examples include read/write requests to retrieve/store data from/to disk drives, utilities to display graphics on the user's screen, network interfaces to facilitate communication with other systems, etc.

---

[10] As you might expect, the architecture of most modern operating systems is far more intricate than that depicted.

There are many advantages to this design. The two most important are that the OS retains control of all system resources, and programmers can remain blissfully ignorant of all the technical tedium associated with managing system components.

The Kernel

The *kernel* serves as the core of any OS. (Hence its name.) It manages all system-level operations and ensures that applications have permission to access any requested service. It is also the component that manages processes, schedules execution, and maintains system integrity.

There are very few system-level operations that are not under the direct control of the *kernel*. It's the part of the OS that loads first during the boot process and manages the completion of system startup tasks. It is also the component that oversees the orderly shutdown of services when you want to power down the system.

Device Drivers

Have you ever considered how many types and models of computers (desktops, laptops, tablets, smart devices, etc.) OSs (like Microsoft Windows, Linux, macOS, and Android) run on? Or that thousands of peripheral devices can interoperate with them? The number of combinations is staggering. Nonetheless, when you buy a new component and plug it into your computer, it just *works*. (Well, usually.)

So, how do OS vendors address this issue? Through the design and use of *device drivers*.

Developed by hardware vendors, *device drivers* usually ship with the purchased component (or, more commonly today, are downloadable via the Web). By design, when you "plugin" a new device, the OS automatically installs the associated driver. At that point, the new component is "visible" on your system and available to all applications. Once installed, *device drivers* have two primary responsibilities: "drive" (i.e., control) the hardware and interact with the OS.

Operating systems invoke device drivers in one of two ways: *directly* or *indirectly*. In the direct case, the OS instructs a driver to perform a specific task. For example, when an application wants to read data from a disk drive (via a *trap*), the kernel will call upon the appropriate device driver to manage the request.

Indirect device driver invocation occurs when the OS responds to an *interrupt*. As mentioned earlier in the chapter, *interrupts* are signals used by hardware devices to indicate that they require servicing. Thus, when the disk drive from the previous example completes the read operation, it will "raise an *interrupt*." In response, the OS will *dispatch* the appropriate driver to service the request.

Software Resources

Although hardware serves as the foundation, not all computing services are "wired." Some system features are software-based. Examples include:

> **Security** Resource accessibility is a "logical" concept managed through a set of permissions implemented in software.

IPC   IPC, or Inter-Process Communication, is a software service that allows processes to exchange data. A common example is *copy-and-paste*. When using any IPC feature, the OS acts as a "telephone company" routing messages among communicating processes.

Allocation   Some resources, like the CPU, require exclusive access. The *kernel* arbitrates among requesting applications to ensure that multiple processes do not use such resources concurrently.

Like their hardware counterparts, applications access software-based services via system calls (i.e., *traps*). Extending the paradigm, system programmers often implement software-based services as *device drivers*.

Memory Management

Have you ever had to rearrange pots in a dishwasher to make room for a few more items? Operating systems deal with a similar issue when managing memory. As processes come and go (execute and terminate), they acquire and relinquish memory segments. As a result, memory can become *fragmented*.

For example, as depicted in Figure 9.5, there is enough memory available in the system to allow the New Application to execute: it requires 3 megabytes, and memory segments 1 and 2 contain 4.5 megabytes collectively. Unfortunately, the segments are not *contiguous*, and neither is large enough individually to meet the need. We refer to this condition as *memory fragmentation*.

Operating systems address fragmentation issues in one of two generic ways: *reactive* and *preventative*. We'll discuss one example of each.

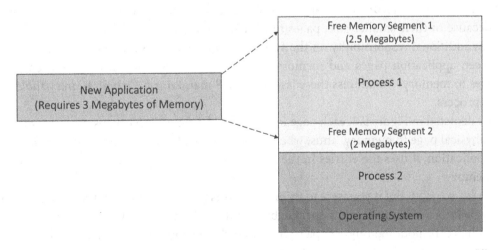

FIGURE 9.5   Fragmented memory.

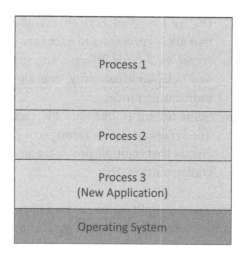

Main Memory

FIGURE 9.6   Memory after compaction.

Let's begin with a *reactive technique* called *compaction*. As illustrated in Figure 9.6, when memory becomes fragmented, the OS can reorganize it such that all the "free" segments coalesce.

Although compaction only occurs when necessary, it has one major disadvantage: all processing must pause while the OS repositions the applications.

A preventative memory management technique, called *paging*, introduces a concept called *virtual addressing*. Let's see how this works.

Instead of one contiguous segment, the OS divides memory into small "chunks" called *pages* (see Figure 9.7). As the OS prepares a program for execution, it divides its process image into page-sized chunks and loads them wherever there are available pages in memory. Please note, as depicted in Figure 9.7, an application's pages need not be contiguous or sequential.

Because of their smaller size, pages minimize fragmentation. However, this approach places additional responsibility on the system: how does the OS maintain the association between application pages and memory pages, and how does it track where each page resides in memory? To address these issues, the OS maintains a *Virtual Address Table* for each process.

As depicted in Figure 9.8, each page in the *Virtual Address Table* maps to a corresponding physical page in memory. Thus, when the system needs to reference a piece of data for an application, it uses the entries in the *Virtual Address Table* to locate the physical page in memory.

So far, so good. But if you were to think about this for a moment, you'd realize that's not enough information to locate a specific byte of data. That is, the system has determined only the corresponding physical page; it still needs to compute the byte's location *within* the page. To determine the exact address of a given byte, the system must calculate its relative offset from the beginning of the *virtual page* and then reference the corresponding location on the *physical page* (in memory).

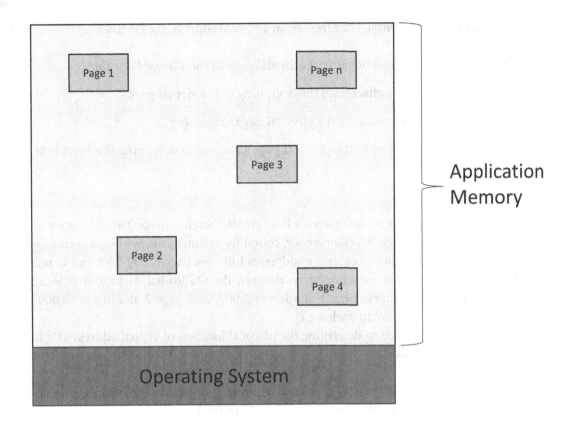

Application
Memory

Operating System

Main Memory

FIGURE 9.7 Paging.

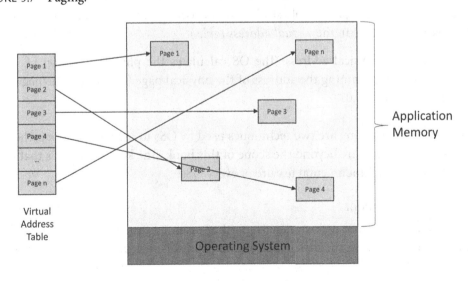

Application
Memory

Virtual
Address
Table

Operating System

Main Memory

FIGURE 9.8 Paging and virtual memory.

In sum, to determine a physical address from a virtual address, the OS must:

- Use the virtual address to identify the virtual page in the *Virtual Page Table*

- Compute the address's offset from the beginning of the virtual page

- Use the virtual page to identify the physical page in memory

- Add the computed offset to the physical page's address to determine the exact byte location in memory

It's time for an example.

Let's say that the page size in our system is 100 bytes.[11] Therefore, page 0 would represent memory addresses 0 through 99. (Remember, computer scientists always begin counting at 0.) Similarly, page 1 would represent the addresses 100–199; page 2, 200–299; and so on.

Let's further assume that for some given process, the OS loaded its page 0 at location 1,000 in physical memory; page 1 at address 5,000, and page 2 at address 10,000. (Remember, pages do not have to coalesce.)

Now, if the system needed to determine the physical location of virtual address 110, it would compute it as follows:[12]

1. Determine the virtual page: The virtual address, 110, resides on virtual page 1. (Remember, page 1 contains addresses from 100 to 199.)

2. Compute the page offset: The virtual address, 110, is offset 10 bytes from the beginning of its page (110 − 100 = 10).

3. Determine the physical page location: Per the above, we know that virtual page 1 resides at address 5,000 in physical memory. (In a running system, the OS would "lookup" this value in the *virtual address table*.)

4. Compute the physical address: The OS calculates the physical location of virtual address 110 by summing the address of the physical page (5,000) and the page offset (10): 5,000 + 10 = 5,010.

Compaction and paging are two techniques used by OSs to manage memory; there are several others, but they are beyond the scope of this book. The key takeaway is that memory management is a fundamental feature of every OS.

File System Management

Managing and organizing disk storage is another essential task required of OSs. Absent centralized control, using the disk could become a hit-or-miss proposition. For example,

---

[11] Page size varies by system; however, because computers use binary arithmetic, they are usually multiples of 512.

[12] For convenience, we're performing these calculations in Base-10. Remember, computers would execute this calculation in binary.

how would we prevent one application from inadvertently (or deliberately!) overwriting another program's data? How would applications know which areas of the disk were available to use? How would programs locate previously stored data?

Thus, to ensure privacy and maintain data integrity, OSs must control this resource. Let's see how.

The Abstraction

Most OSs organize disk storage using abstractions familiar to users.[13] One of the most common approaches mimics the way organizations historically managed paper: files and folders. Data resides in *files*; files reside in *folders*.[14] Most modern OSs extend this analogy and allow folders to contain other folders (called *subfolders*), creating a hierarchy.

For example, you could create a folder named TAXES and save all your IRS-related files in it. However, over time, you might find it problematic to locate all the files associated with a specific tax year. Understanding that folders may contain other folders, you might opt to reorganize your tax records in a structure like that depicted in Figure 9.9.

As the figure demonstrates, you can create a custom folder *hierarchy* to meet your specific needs. In this case, we have a *parent folder* called TAXES that contains *subfolders* named 2019, 2020, and 2021. Note that, as highlighted in the subfolder 2021, folders may include both files and other subfolders.

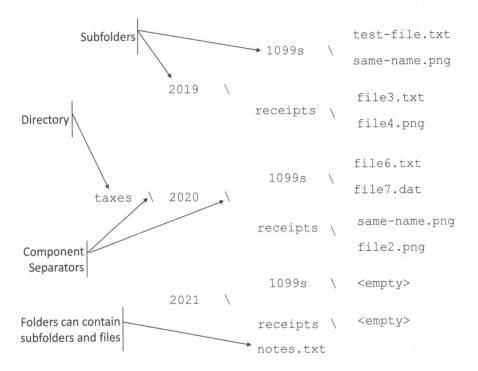

FIGURE 9.9   Example of tax folder hierarchy.

---

[13] Some applications, like databases, prefer using "raw" disk and bypass the file system.
[14] Some systems (e.g., Linux) call folders *directories*.

One syntactical consideration: The backslash character ("\")[15] does not appear in the filesystem or on disk. It's for notational convenience and serves to separate components, allowing us to reference a file by its *pathname*.

For example, in Figure 9.9, the *pathname* for the file TEST-FILE.TXT is as follows:

TAXES\2019\1099s\TEST-FILE.TXT

As might be obvious, file names within a folder must be unique. (Otherwise, how could we distinguish them?) However, we can name two (or more files) identically if they reside in different folders as in:

TAXES\2019\1099s\SAME-NAME.PNG

and

TAXES\2020\1099s\SAME-NAME.PNG

Note that their *pathnames* uniquely identify each file even though they share a common file name (SAME-NAME.PNG).

The OS implements the file/folder abstraction on your behalf; thus, you typically don't need to concern yourself with the details. For example, regardless of the underlying disk technology, the file system on your ten-year-old laptop will appear to function identically to the one that shipped with your brand-new tablet.

Nonetheless, let's take a peek under the hood.

The Implementation

File system implementation varies based on the OS, its version (the vendor might deliver improvements over time), and the underlying disk technology. Nonetheless, we can present a general understanding of how OSs manage disk. In other words, another *abstraction*.

As with memory, OSs divide disk storage into "chunks." However, in this case, we call the chunks *blocks*.

Blocks serve as the basic units of storage (i.e., the smallest "chunk" of disk allocated to a file) and transfer (i.e., reading and writing from/to disk). They can vary in size from 512 bytes to 5,120 bytes (or larger) depending on the underlying disk technology and the system's design specifications.

For example, let's say the system we're working on uses a block size of 512 bytes, and you create a file that contains just one sentence:

THIS IS A TEST.

From the initial capital letter ("T") to the ending period ("."), the sentence contains a total of 15 characters (the three intervening blank characters count). Nonetheless, the file system will allocate a full block of storage (in this case, 512 bytes) and assign it to the file.[16]

---

[15] Some file systems (e.g., Linux) use the forward slash ("/") character as the component separator.

[16] We refer to the unused portion of the block as *internal fragmentation*. Although it may appear wasteful, there are sound technical reasons for this design approach.

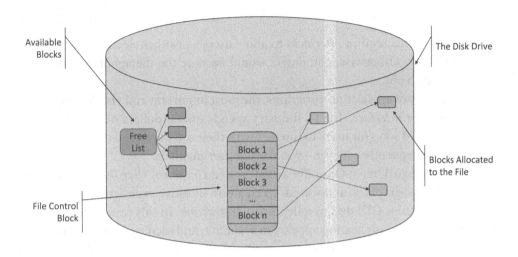

FIGURE 9.10   File system state.

Moreover, if you were to edit this file later, the file system must copy the entire block into memory, not just the 15 bytes in question.

Block *allocation* is dynamic: as you add data to a file, the system will obtain unused blocks from the *free list* and append them to the file. When you delete an entire file or just remove some data from it, the system will *deallocate* (remove) unused blocks from the file and return them to the *free list*.[17]

Let's examine how this process works.

Figure 9.10 depicts the current state of a conceptual file system. Located on the left side of the diagram is the *free list*, which maintains the inventory of all currently unassigned disk blocks. When a file grows (or is created), the file system obtains (removes) one or more blocks (as required) from the *free list* and adds (*allocates*) them to the file. When a file shrinks (or is deleted), the file system removes (*deallocates*) blocks from the file and returns them to the *free list*.

Near the center of the diagram is a special block called the *file control block*.[18] It contains data about a file, such as its size, type, and creation date.[19] It also includes "pointers" to the data blocks currently allocated to the file.

Please note that, like memory pages, blocks allocated to a file need not be sequential or contiguous. Moreover, we compute data addresses similar to the approach we used for paging. Thus, byte number 512 of a file is the zeroth byte of block number 1. (Remember, we count from zero in the world of computers.)

File system management is among the key features provided by OSs. It ensures the integrity and reliability of the most valuable resource residing on any computer: *data*.

---

[17] Nota Bene: The file system does NOT erase data on blocks it returns to the *free list*. The data is still readable if you know how to do it. Refer to the section below entitled, *Security*, for more details.

[18] Remember this is a logical representation. Not all file systems call this a "control block."

[19] We often refer to this type of data as *metadata*. Technically, metadata is data about other data. For example, the bytes contained in a file are its *data*; however, an attribute like "file size" is *metadata* in that it *describes* the data in the file.

User Interfaces

All computer systems require *interfaces* to allow users—or other electronic devices[20]—to interact with them. Otherwise, computers would become the metaphorical "black box:" No data in, no data out.

For the majority of application programs, the most important and visible aspect of their design is their *user interface*. The IT industry spends untold dollars developing intuitive and straightforward ways for users to interact with their software. And they're succeeding. Consider how infrequently—if ever—you consult user manuals.

To support that goal, most OSs offer an integrated *Graphical User Interface* (or GUI).[21] GUIs employ icons, graphics, and sounds to create an intuitive experience for users.

As with file systems, GUI design relies on an *abstraction*. In this case, the metaphor is a desktop. Each running application appears in a *window*, and each window is akin to a piece of paper on a desk. As you are free to move and reshuffle documents on a physical surface, so can you reposition and reshuffle windows on your screen.

As a rule, GUI interfaces divide responsibilities between the OS—as implemented by a *Window Manager*—and individual applications. The primary task of the *Window Manager* is to maintain the desktop. Based on mouse clicks or screen touches, it positions windows and determines which parts of any given window are visible.

Applications are responsible for maintaining the contents of their windows. Obviously, as data changes, programs must update their displays. However, what might not be so obvious is that applications must also "repaint" their windows whenever one of them becomes uncovered (*exposed*). Figure 9.11 provides an example.

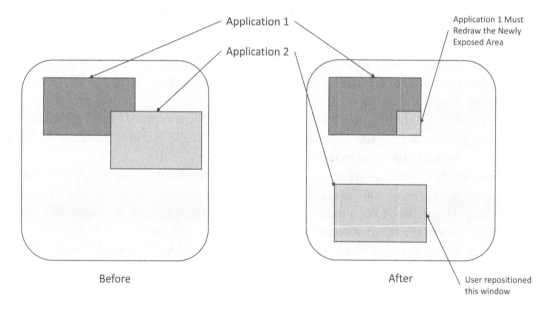

Application 1

Application 2

Application 1 Must Redraw the Newly Exposed Area

Before

After

User repositioned this window

FIGURE 9.11   Window repainting.

---

[20] We usually refer to a system-to-system interface as an API, or Application Programming Interface.
[21] With some operating systems, the GUI interface is an add-on product.

In the "before" state, Application 2's window covers (in GUI parlance, *occludes*) the lower right-hand corner of Application 1's window. When the user repositions Application 2's window (the "after" state), the lower right-hand corner of Application 1's window becomes visible. When this occurs, the Window Manager sends Application 1 a "repaint" request. Upon receipt, Application 1 responds by redrawing its window (or at least the newly exposed portion).

Graphical interfaces have become the norm. Indeed, most users don't give them a second thought: they just launch applications and interact with them intuitively. But, like every other aspect of modern computers, it's the OS that provides the foundation for their success and ease of use.

System Initialization (Booting)

Obviously, when your computer is powered off, nothing is running—including the OS. So, when you apply power, how does the OS *come to life*? As it turns out, there is an intricate orchestration of events that we call *system initialization* or the *boot process* that initiates the OS.

One note before we continue: Every system employs a unique boot process. Some are more elaborate than others based on need and applicability. For example, you might not mind waiting 45 seconds for a desktop computer to boot. However, most people would consider that an eternity for a smartphone. Thus, as with many of our prior discussions, the material that follows describes the startup process in conceptual terms.

For a typical computer (desktops, laptops, and some tablets), the boot process unfolds in stages. When we apply power to a system, a small program, called the *bootstrap loader*, begins running. This program ships with the hardware and performs one task: it executes the *stage one loader* of the OS that runs your device.

Although larger than the bootstrap loader, the stage one loader has only one task as well: it executes the *stage two loader*. (In some systems, the stage two loader might be a tiny OS in its own right.) Unlike the programs that preceded it, the stage two loader performs several tasks, including initializing memory, loading the primary OS, and initiating system execution.

As part of its startup tasks, the OS verifies the viability of all hardware components, loads device drivers, and initializes the environment. The boot process completes when the OS is ready to execute user applications.

When you start a system by applying power, it's considered a *hard boot*. A *soft boot* occurs when you instruct the OS to restart via software. (This typically happens after a system upgrade. The OS may inform you that it must reboot the system to complete the installation process.) For many systems, the soft boot process may not require as many steps as a hard boot.

To control software versions and ensure security, some organizations boot their computers via a network. That is, the boot loaders and the OS reside on remote servers. In such cases, it's likely that the system will not boot without a functioning network connection.

One final note: Many computers allow you to choose which OS you want to launch. For example, many users run both MS Windows and Linux on their computers. In such cases, the system prompts you to select the OS you want to run when you power up.

Security

In addition to providing a stable, robust execution platform, most OSs also enforce security policies. This task includes managing users, safeguarding system resources, and protecting running processes from other misbehaving or nefarious programs.

System security is a multifaceted responsibility; let's begin with users.

User Security

When it comes to security, an OS's highest priority is to ensure that only authorized users may access system resources. To that end, the OS will impose an *authentication process* that users must undergo before it grants them admittance. This procedure can range from a simple login and password combination to employing elaborate biometric verification techniques.

Authenticating users has many benefits, including:

Identification Typically, each user is assigned a unique id that the OS uses to track resource usage. Thus, system administrators (and device owners) "know" what each user does on their systems.

Authorization Resource entitlement need not be universal. For example, consider systems that support a health-care facility. Clearly, billing personnel should not have access to patient medical histories. In such cases, system administrators can selectively impose restrictions on data access.

Auditing The system can record every command (even every keystroke if required) a user enters. This type of tracking is of particular concern in secure environments like government installations. However, this might also be an important feature used in legal, medical, and laboratory environments.

Monitoring Systems can track resource consumption by user. Such data collection supports billing processes and alerts administrators when individual users consume resources in excess of their entitlements.

Process Security

On most systems, an executing program assumes the privileges of the user that invoked it. Consequently, the process can only access resources authorized for that user. For example, two users could launch MS Word on the same system. However, one user might be able to edit a file that the other user can't. In most cases, it's the OS that enforces such restrictions.[22]

---

[22] This might not hold true for all resources. In some cases, application software—like a database management system—might determine authorization.

The OS also ensures that executing processes don't violate system constraints. For example, earlier in this chapter, we noted that addresses are just numerical values. Thus, what's to stop a process from trying to read (or write to) a portion of memory it doesn't control? It's usually the OS that prevents such erroneous or pernicious behavior.

Data Security

We'll discuss the topic of data security in detail later in the text. However, before we complete this section, I want to alert you to an issue of which many computer users remain ignorant.

Earlier in this chapter, we observed that when you delete a file, the system returns the now-unused disk blocks to the free list. However, and please take note, **the OS doesn't erase anything**. Any information contained on those blocks is still there. That's why you may "undo" a file deletion on some systems—the data remains unaltered on the deallocated blocks.

As a result of this design, hackers, computer forensic investigators, and bored teenagers with time on their hands can read "deleted" data if they have access to your system. Consequently, you run the risk that financial accounts, social security numbers, and your aunt's secret apple pie recipe might fall into the wrong hands.

To thwart this threat, you can use one of the many commercial tools that permanently erase data in deleted files. If protecting personal data is a concern to you, I suggest investing in one of these products.

Common Services and Utilities

In addition to the services discussed above, most OSs provide many other features and utilities. I say "most" because some OSs that control small devices—like those in smartwatches—might not offer as many services as their big siblings, which manage desktops, laptops, and tablets.

The sections below examine some of the most common services and utilities provided by OSs.

File Explorers

Even the most sophisticated file system would be useless if it didn't allow users to review and manage their files and folders. All PC OSs provide such tools. I'm sure you're familiar with *File Explorer* for Windows or *Finder* that ships with macOS.[23] Both utilities offer graphical representations of their respective file systems and allow users to open, create, delete, and copy files and folders.

Disk Administration

Over time, files and folders can become fragmented or corrupted. Thus, most OSs ship with disk administration tools that repair such anomalies. Many systems also include utilities that optimize file structures to increase performance.

---

[23] File system access is slightly different under operating systems like Linux.

Backup and Restore

All digital devices fail. And let me be clear on this: it's not a case of *if*, but *when*. Disk drives are no exception to that rule. As a result, most OSs include backup utilities that copy files from one disk drive (called the *primary*) to another (called the *secondary* or *backup*).

Because of advances in manufacturing processes, disk technology has become so reliable that many users have become complacent and don't backup their systems regularly. Don't be one of them. If your data is important enough to save on disk, it's valuable enough to backup as well.

Disk drives have become very inexpensive and are simple to install. Buy one. Better yet, buy two. I know it's easy for me to spend your money—but consider what the cost to you might be if your primary disk drive failed and you couldn't recover your files and data.

If you don't like the backup tool provided with your system, there are many inexpensive (read *free*) offerings from reliable third-party vendors. Use one of them regularly.

Clipboard

One of the most useful features included in modern computing environments is *copy-and-paste*. The ability to grab data from one application and insert it into another minimizes human error and saves us from the tedium of retyping information. However, as should be evident by now, such processing is not magic. It's merely a set of conventions[24] that programs must adopt to participate in this processing.

Network Administration

Over the past two decades or so, networks have become extremely reliable and easy to use. However, that's not to suggest that nothing ever goes awry: components fail, software has bugs, circuits experience outages, and so on.

To help address such anomalies, OSs provide tools to analyze, diagnose, and sometimes repair network problems. (Obviously, software tools cannot fix a hardware failure; you must physically replace the defective component.)

Most systems also provide tools that display your computer's IP and MAC addresses, evaluate your network's speed, and determine whether a device your system is communicating with is responding.[25]

EXAMPLE: DISK READ SCENARIO

Okay, it's time for an example. Let's start with the scenario: a program (like a word processor) needs to read data from a file to display it on the user's screen. What follows is a summary of the OS tasks required to retrieve the blocks from external storage and make the data available in memory to the requesting process.

As you may recall, processes use *system calls* to *trap* into the OS to request a service. In this example, the word processor would invoke a *read* function to retrieve a data block from a file.

---

[24] We usually refer to conventions as *standards* in the IT world.
[25] We call this a *ping* test.

The OS receives the request at its *application interface layer*, which, in turn, forwards it to the *kernel*. The kernel verifies that the requesting process has permission to read the requested data block, *suspends* the requesting process (until the read operation completes), and forwards the request onto the associated *device driver*. After the device driver initiates the request with the disk's controller, the kernel *schedules* the process with the highest priority to resume execution.

System processing continues until sometime later when the disk completes the read request, and its controller raises an *interrupt* with the OS. In response, the kernel dispatches the appropriate *device driver* to service the disk drive.

At this point, the data block requested by the program resides in a memory location controlled by the kernel; that is, the application cannot access the data directly. Thus, the kernel must copy the data from protected memory into a location accessible by the application and change the process's state to *runnable*. Eventually, based on system priorities, the OS will schedule the program to run, at which point it can process the data.

Admittedly, completing such a request is slightly more complicated than just described. However, most of the intricacies are in the details. For example, the OS contains a substantial amount of code dealing with errors and exceptions that can occur during this scenario.

ADVANCED TOPIC: TYPES OF OPERATING SYSTEMS

The tired trope, *one size does not fit all*, holds true even for OSs. Consider that the OS running a smartwatch will contain a feature set that varies widely from that of an OS managing a laptop, a smart car, or one that guides rockets to the International Space Station.

The sections that follow present an overview of some of the most common types of OSs.

Single-Tasking vs. Multitasking

As mentioned previously, a computer with a single CPU can execute only one instruction at any moment in time. In line with that, a *single-tasking*[26] OS can only run one process at a time. Though that might appear limited, such systems are still widely used today in devices like fitness trackers and intelligent thermostats.

More commonly, however, OSs are *multitasking*. That is, the OS can manage the execution of multiple processes concurrently. The question is: How?

As we discovered earlier in this chapter, the CPU is just another resource like disk or memory. But instead of using a first-come, first-served methodology that might work for some peripheral devices, the OS manages access to the CPU using a technique called *time-sharing*. Under this approach, each process receives a *time quantum* during which it has exclusive access to the CPU. When its allotted time expires, the OS suspends that process and schedules another one to run.

For example, assume the time quantum for a given system is one second. When some process, say P1, begins executing, the OS sets a timer (using yet another resource: the system clock). When the timer expires (after one second in this example), the OS resumes

---

[26] In this context, the term *task* is synonymous with *process*.

control of the system (via a *clock interrupt*) and performs a *context switch*, pausing process P1 and scheduling another process, say P2, to start executing.

But clock interrupts are not the only way programs surrender control of the CPU under a timesharing OS. Because disk and network operations are ponderously slow compared to CPU speeds, processes also relinquish the CPU whenever they request services from peripheral devices.

In modern systems, *Context Switching* occurs so smoothly that human users only become aware of it when the system is under intense load and begins to appear "sluggish." Otherwise, multiple applications seem to run simultaneously.

Single-User vs. Multi-User

Single-user OSs, such as those that run on most PCs, grant access to only one user at a time. In contrast, a *multi-user* OS—such as Linux—grants multiple users concurrent access to the system. That is, unbeknownst to each other,[27] authorized users can interact with the system simultaneously without affecting anyone else. Note that by their very nature, multi-user OSs must also be multitasking.

As you might expect, supporting concurrent users places additional burdens on the OS. It must ensure that users can't affect each other, can't access each other's confidential data, or consume a disproportionate amount of system resources.

Real-Time Operating Systems

We are most familiar with the OSs that allow processes to execute until their time quantum expires or request a service from a system resource (e.g., a disk read). Thus, at some point, every executing application will eventually relinquish control of the CPU, allowing other programs an opportunity to run.

However, this is not always the desired scenario. There are times when a given process is so critical that the OS should not suspend its execution.

For example, consider the OS controlling the launch of a Falcon 9 rocket. If an alarm occurs signaling that a dangerous condition has occurred, the OS should never interrupt the process attempting to rectify the problem. That could spell disaster for the astronauts.

A *real-time* OS ensures that the highest-priority task remains in control of the processor. Only a higher priority task can interrupt the execution of the one currently running.

SUMMARY

We've covered a considerable amount of material in this chapter, including how OSs manage processes, resources, program execution, and security. We also discussed the boot process and how systems expose their services to human users using abstractions.

In Chapter 10, we'll expand on this knowledge and discover how developers design and build modern software applications.

---

[27] In practice, there are ways for a user to determine who else is using the system.

Software Architecture and Design

You can mass-produce hardware; you cannot mass-produce software—you cannot mass-produce the human mind.

MICHIO KAKU

INTRODUCTION

Throughout this text, we've used the term *application* rather loosely. We described it as a set of instructions that perform a specific task. But how do applications fit into the conceptual picture of computer systems we've been painting? Before we get too technical, let's add some more brushstrokes to our portrait.

As noted previously, the combination of hardware and operating system provides a platform upon which applications execute. And we've also compared the CPU to the human brain and the operating system to the human mind. So, where do applications fit into this metaphor?

We can envision applications as learned behaviors. Repetitive tasks that we can call upon whenever we need them but can adapt when necessary. For example, consider a pianist that knows how to play a particular tune. Obviously, the player could repeatedly regale us with the identical rendition of that song. But what if we asked that musician to play the piece slower? Or, in a different style? The pianist would have to diverge from the memorized version to honor either request. Another way to view that example is that the player adapted to new "input data."

A similar type of adaptive processing occurs with computer applications as well. Think about a simple calculator program: it "knows" how to add, but its results vary based on the numbers we input into it. With this conceptual interpretation in mind, let's formalize our understanding of applications.

WHAT IS A COMPUTER APPLICATION?

Previously, we've characterized a computer application as *a collection of one or more programs designed to serve a specific need*. Let's refine that definition a bit.

> An *application* is a finite arrangement of instructions packaged as one or more programs that, when executed, use data to accomplish a specific, targeted set of tasks within a computing environment.

Let's begin analyzing that definition by scrutinizing the phrase: "one or more programs." Most computer users tend to use the terms "program" and "application" interchangeably. However, in the world of IT, the words have distinct meanings. A *program* is a single set of instructions often contained in a single executable file.[1] Whereas an *application* comprises multiple programs, each of which resides in separate files.[2]

Let's move on to the phrase "targeted set of tasks." All applications limit their functionality to some degree. For example, some digital music players support only one type of song file (e.g., .wav). Other applications, like a sophisticated word processor, may integrate many related features such as a dictionary, a grammar analyzer, and a spell checker. Nonetheless, despite variance in scope, every application addresses a specific set of needs. They are not "all things to all people."

The phrase "use data" in the definition is deliberately vague. Applications may operate on any digital information (e.g., text files, audio streams, numeric data, biometric input, etc.) in any combination as needed. An application's data may reside locally, on a remote server, or in a database connected via a network—in any combination. Note that some applications might also generate data. For example, consider a Sudoku game app that might use programmatically generated random numbers[3] to create each puzzle.

Though not part of the definition, let's take a moment to dispel any confusion between "app" and "application." In a formal sense, there is no distinction between the two terms. However, in practice, we tend to say "application" when referencing programs running on desktop and laptop computers and use "app" when describing programs that run on mobile devices.

WHERE ARE APPLICATIONS USED?

Unless you've lived in solitude for the last fifty years, the answer to the question, "Where are applications used?" is obvious: everywhere. We should probably end this section right here. However, for the sake of completeness, I'd like to discuss some uses of software that might not be so obvious.

Business

Most of you likely use software applications at your workplace: email programs, word processors, spreadsheets, payroll systems, etc. In terms of direct use, enough said. However, you might be unaware of some indirect interaction you might have with computer applications.

---

[1] Though not 100% accurate, this definition will serve our needs. Specifically, there are execution environments wherein individual programs may comprise separate files.

[2] Unfortunately, even IT professionals commonly refer to a single program as an application.

[3] Technically, in computer science, we call these *pseudorandom numbers* because, although they are statistically random, they repeat their sequences when initiated from the same starting point.

For example, some logistics companies track their vehicles using satellite technology (often without the operator's knowledge). Thus, an organization can monitor how many hours its drivers are behind the wheel each day and receive real-time alerts when one of its vehicles deviates from its assigned route.

Another example is when you buy a garment in a department store. When the clerk "rings you up," the register might send the purchase information to a system that tracks inventory and automatically initiates replenishment orders with the vendor.

An extreme case of automated ordering is *just-in-time delivery*. It costs money when inventory sits idle on a shelf. To reduce overhead, companies buy goods only when they need them. For example, a car manufacturer's inventory system can compute how many vehicles the automaker will build tomorrow and automatically order the appropriate number of tires to arrive at the factory's loading dock in the morning. Pretty slick.

Music

We've already discussed how applications process digital sound via audio files and streaming services. However, what you might not know is that software generates much of the music you listen to today. And I'm not just referring to the "weird" sounds that are "otherworldly" and clearly artificial. Familiar instruments such as violins, horns, drums, organs, and pianos are not always performed "live." In many cases, composers use Digital Audio Workstations (or DAWs) to create those sounds using software (and sometimes specialized hardware).

As an aside, digitized music is one example of how software has far-reaching and often unintended consequences. DAWs have democratized the world of music and liberated musicians. Anyone with a laptop can create, package, and market music without the need of publishing companies or record labels. As a result, we can listen to many artists that, in the past, we might have never known existed. Of course, that also means that we might have to wade through a flood of junk from "artists" who have only a fleeting relationship with talent. (Alas, as with most things in life, we must take the good with the bad.)

Education

In recent years, distance learning—where students are physically remote and interact with their instructors via computer[4]—has become commonplace. However, application software is far more integrated into pedagogy than just connecting pupils and teachers.

Although it might not be obvious, many online courses adjust lessons based on each student's progress and test results. When implemented correctly, this is self-paced, one-on-one education at its best. Many publishers can also custom-tailor textbooks to align precisely with the syllabus of individual courses offered at various universities. Thus, the contents and sequence of topics for the same book can vary by school.

Educators also use advanced graphics to develop visual images that facilitate the presentation of complex structures and ideas. For example, medical schools create three-dimensional models of complex organs like the heart and lungs that allow students to "travel through" them. Very cool.

---

[4] This became quite prevalent during the COVID epidemic of 2020.

Health Care

In the world of health care, software performs much of the diagnostic testing we undergo. For example, with the aid of sophisticated computer graphics, MRIs, CAT scans, and sonograms create detailed, three-dimensional images of our internal organs.

But medical software does so much more. It aids doctors in diagnosing illnesses. As an illustration, recent studies have demonstrated that AI software may be more accurate in identifying disorders of the eyes than ophthalmologists. Moreover, it appears that AI technologies may also aid in developing new therapeutic drugs and vaccines and improving the early diagnosis of progressive neurological diseases such as Alzheimer's and Parkinson's.

Military

Much of what the military does with technology is unknown to the general public (that includes me). However, I think we can state without reservation that software drives the modern army. It coordinates battles, repositions personnel, directs smart weapons, and manages encrypted communication.

Moreover, battlefields have become increasingly digital. Armies in the traditional sense are becoming obsolete, replaced by "techies" who, without leaving the comfort of their favorite armchair, fight on land (using drones and smart tanks), sea (directing remote-controlled watercraft), and air (controlling pilotless fighter jets).

More alarmingly, however, due to technology's influence, the perception of the battle-field has morphed. Today, "techie" combatants fight for more than traditional military objectives—they grapple for control over private-sector digital resources as well. For example, foreign hackers might attack a country's banking systems to undermine its financial stability. Or target a region's transportation infrastructure to wreak havoc on roads and debilitate commerce.

THE DESIGN OF SOFTWARE APPLICATIONS

Most of us assume professionals review and analyze problems before deciding how to address them. For instance, we would expect that plans for a four-bedroom center-hall colonial will differ significantly from that of a ten-story apartment building. And we would not be shocked to discover that the designs for a three-hundred-seat jumbo jet would not have much in common with that of a modern, sophisticated stealth fighter.

Yet, it's been my experience that to anyone who hasn't developed software, writing computer applications seems amazingly simple:[5] List out the instructions, then execute the program. Easy as pie, right?

Alas, no.

Like any other engineering discipline, developing software applications of any appreciable size requires serious forethought. In the IT world, we divide that planning into two phases: *architecture* and *design*.

---

[5] This has been the assumption of many of my clients over the years.

Let's begin with the definitions. First, *architecture*:

Software architecture defines the major components (also called modules) that comprise an application. It also establishes how these components interact.

Now, *design*:

Software design is the process of creating specifications for modules that adhere to architectural requirements.

Before we continue, I should note that there's an ongoing debate among IT professionals regarding the distinction between architecture and design. Specifically, the argument rages over where architecture ends and design begins. Clearly, there's some overlap and flexibility in the above definitions.

For example, a Software Architect (the IT professional who develops system architectures) might decide—for sound reasons—to divide a large module into several smaller ones. This architectural choice clearly affects the design because there are more components and interactions to address.

That said, we'll ignore the academic subtleties and describe the distinction in practical terms.

To help delineate the boundary between software architecture and design, we can draw upon an example from the construction trade. If we wanted to build a house, the first thing we'd need is its blueprints. These drawings specify the sizes of rooms, heights of ceilings, locations of windows, etc. Unmistakably, this is architecture, and we typically acquire such plans from an architect.

However, although the blueprints would also identify some required subsystems such as the furnace, water heater, and kitchen appliances, it would not include those components' designs. Each unit we install in our house would be designed separately.

With that model in mind, let's see how this works with software.

SOFTWARE ARCHITECTURE

We can view application software "construction" in much the same way as our house example from the previous section: layout the significant components and develop individual designs for each. Unfortunately, it's a tad more complicated than that. To make the point, let's continue with our house construction example for a moment.

Consider that there are many categories of buildings. To begin a project, we'd need to choose the type of structure: office complex, college dormitory, two-family home, single-family home, etc. Then we'd need to select the model. For example, if we were building a one-family dwelling, our options might include side-hall colonial, ranch, split-level, etc.

Similarly, there are many "styles" of software architecture. You may have heard of some of them: peer-to-peer, monolithic, three-tiered, layered, domain-driven, and so on. Each type (or *pattern*) serves a particular need. And, like the construction industry, we'd have to select one at the onset of a project.

To demonstrate how IT professionals build applications, we'll introduce one of the most widely used software architectures.

MVC

Let's kick off our architectural discussion by considering how most applications function. They begin execution by retrieving data from a repository of some sort and then displaying it to the user. Next, they allow users to manipulate the displayed data in predefined and approved ways. Finally, they record and save any changes for later use.

Although simplified, this description illustrates how typical user-oriented applications operate.[6] What follows are some examples of systems that adhere to this paradigm.

Banking	After you log in, a typical banking app displays your current account information retrieved from the bank's database. It then allows you to execute a limited set of transactions (e.g., deposit checks, transfer money, pay bills, change your telephone number, etc.). Before it exits, the app saves any changes you make in the bank's database.
Shopping	When you begin a session, a shopping app displays the vendor's products for your consideration. If you want to buy any, it allows you to "place" them in your "shopping cart." When you "checkout," it records any purchases in the vendor's database. The next time you log in, you can check the status of your order.
Stock Trading	Most stock trading apps allow you to view your portfolio and the current state of the market. They also allow you to execute trades that adhere to the SEC [7]policies and record the transactions in your account.

One of the more common software architectures that support these types of applications is Model-View-Controller (MVC).

Model-View-Controller (MVC) is a software architectural model comprising three significant subcomponents: the *Model* (the data subcomponent), the *View* (the user interface subcomponent), and the *Controller* (the subcomponent that updates the *View* and maintains the integrity of the *Model*).

As implied by its definition, MVC partitions application functionality as follows:[8]

- The *Model* is the database (i.e., the location and structure of application data).

- The *View* extracts data from the *Model* (via the *Controller*) and presents it to the user; it also delivers all user-initiated modifications to the *Controller* to update the *Model*.

---

[6] This is not to say there aren't many other common architectures for user-based applications.

[7] The Security and Exchange Commission (SEC) is the government agency responsible for establishing and enforcing market regulations.

[8] Don't worry if the MVC components seem out of sequence; as we will see, they all work in unison.

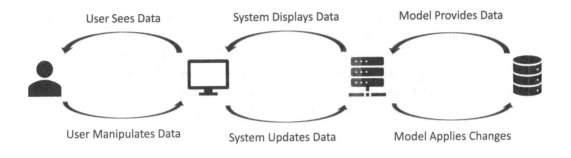

FIGURE 10.1   MVC component interaction.

- The *Controller* is the component that protects the integrity of the *Model* and manages all changes to the data.

As depicted in Figure 10.1, MVC architectural components form a complete solution. The sections that follow discuss each MVC subcomponent in detail.

The View

As you might have inferred, the *View* is the face of the application. It acquires data from the *Model* (the database) and displays it in an appropriate format: text, images (e.g., pictures), graphics (e.g., a pie chart), sounds (e.g., music), lists (e.g., a spreadsheet or a table), etc. The objective of the *View* is to deliver information in an appropriate, meaningful, and user-friendly manner.

As an example, consider an application that automates inventory management for a supermarket. The *View* might present sales data in tabular form for the Produce Manager, itemizing purchases sorted by product. In contrast, the Store Manager might need that same data presented as a bar chart to reveal sales trends. The important point is that applications may support multiple *Views*, each of which displays data in a way that meets an individual user's needs.

The Controller

The primary obligation of the *Controller* is to maintain the integrity of the data *Model*. It does this in two ways.

First, based on their roles and permissions,[9] the *Controller* determines what data each user may see. To illustrate this, let's return to our grocery store example. The database (i.e., the *Model*) for such a system would contain records for every product received into inventory and every item purchased by customers. At any moment in time, it accurately reflects how many packages of a particular brand of mushroom are in stock, their wholesale cost, retail price, etc. Despite the range of available data, the *Controller* might allow Produce

---

[9] Please recall the discussion in Chapter 9 regarding user authorization.

Managers access to only inventory data. By comparison, it may permit the Store Manager to view price and cost information.

The second way the *Controller* maintains the *Model*'s integrity is by managing who has permission to modify data. For example, Produce Managers might have the authority to change order data, but Cashiers would not. However, Cashiers, via their registers, would automatically decrease inventory quantities as they checkout customers. (Note that in this case, the cash register serves as another type of *View*.)

The Model

As discussed earlier, users only see data as presented by *Views*, not in its raw form as it resides on disk. As a result, the *Model* can often be the most challenging component for non-IT professionals to conceptualize because applications manage data in ways that are not readily apparent. That said, let's see if we can bring this into focus a bit.

Most of you are familiar with sound files—almost everyone plays or streams digital music.[10] It's also likely that most of you have encountered a situation where your music player cannot play a song because it's in an unknown or unexpected format. For example, MP3 players cannot play songs contained in .WAV files. That's because the format (i.e., the *Model*) is incompatible with what the player (the *View*) expects. Thus, you might have two copies of the same song (i.e., two different *Models*), but each is only suitable for a specific player (the *View*).

Software developers make similar low-level organizational and formatting decisions when designing the *Model*. Whether they opt to store all application data in one file, multiple files, or a high-powered database management system, designers must determine the most effective way to structure the *Model* to benefit every *View*. (We'll return to this topic in Chapter 11.)

MVC in Practice

When a construction architect designs a center-hall colonial, it's with the expectation that the building is a single cohesive unit. That is, every room is part of the same physical structure. Software architecture differs slightly because designers can package MVC components as a single program or as separate entities.[11]

Let's begin with the simple case—a single program.

MVC: Single Program

One of the most common ways to develop a software application is to place all the code in a single file. Despite this approach, designers can still create programs that adhere to the tenets of the MVC architecture. See Figure 10.2 for an example.

Even though all the code resides in a single file, Figure 10.2 demonstrates that each MVC component remains an independent entity. That is, developers would not include any *View* code in the *Controller* and *Model* sections. Similarly, there is no *Model* code in the *View* or

---

[10] Please review the material presented in Chapter 3 if you have any questions.

[11] As we'll see, separation facilitates software reuse.

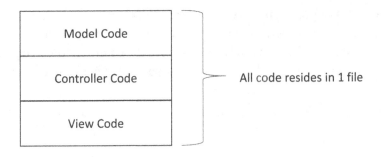

FIGURE 10.2   Single program MVC.

Controller areas. Notwithstanding its apparent simplicity, this approach might not be the best architectural choice when developing large-scale applications.

Let's see why.

MVC: Distributed Components

As alluded to above, it's not practical to develop large applications using just one program file. In practice, IT professionals segment code into multiple files and group them based on functionality. For example, one set of files could contain all the *View* code, another group the *Controller* code, and the remaining files would host the *Model* code.

We call this approach *Modular Design*:

> *Modular Design* is a design technique wherein developers decompose functionality into cohesive, independent components, called *modules*. Each *module* contains all the code required to perform one feature of the system.

The benefits of *Modular Design* are many:

- Multiple programmers can work simultaneously on the same application (by allocating work based on modules).

- Experts can focus on specific areas. As an example, user-interface design specialists can concentrate on the *View*, while database mavens can design and implement the *Model*.

- Code changes can occur in isolation. Consider that if there's a bug in the *View*, it need not affect any code in the *Controller* or *Model* components.[12]

There is another crucial benefit to *Modular Design*. As pointed out in the prior section, if a program runs as a single unit, then all its code and data must reside on the host system. Thus, if multiple users want to run the application, each will need an individual copy of the programs and data loaded onto their computers.

---

[12] Unfortunately, this is not always the case. Sometimes changing code in one module does affect another.

This approach has a potential nasty side effect: if USER1 modifies a value, the other users won't see it. That's because there are multiple copies of the data, and the update took place *only* on USER1's computer. The other users will not see the change in their copy of the data.[13]

This issue brings to mind an old programming parable. It goes something like this:

A person wearing one watch always knows what time it is; anyone wearing two watches is never really sure.

By extension, if there are multiple instances (copies) of a data set and each contains different values, which one is correct?

To address this issue, system architects can design application components such that each can run on separate computers, allowing users to share module instances rather than needing individual copies.

Figure 10.3 provides an example of this approach. As highlighted in the diagram, all MVC components run on separate servers and communicate via network connections.

By the way, there's no need to stop there. We can also allow multiple users access to the same system concurrently. See Figure 10.4 for an example.

Although their *Views* might differ, every user represented in Figure 10.4 shares the same database instance. Thus, any changes by one user are visible to all users. The *Controller* maintains the integrity of the *Model* by ensuring that each user may perform only authorized tasks. Also, because they communicate via network connections, each MVC component can reside on any server in any data center located anywhere on the planet.[14]

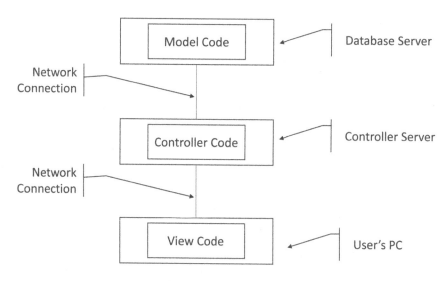

FIGURE 10.3 Distributing MVC components.

---

[13] For some types of applications, this is not an issue. It's only a concern when multiple users want to share data.
[14] Although this statement is true, there are some practical considerations designers must consider: performance, costs, security concerns, etc.

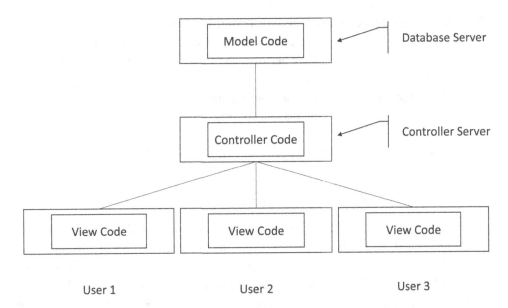

FIGURE 10.4   MVC with multiple users.

This type of distributed MVC is more common than you might think. Indeed, most browser-based applications and phone apps rely on some variant of this architecture.

MVC: Advantages and Disadvantages

MVC boasts numerous advantages as compared with single-component (monolithic) designs, including:

Component Reuse   When designed correctly, MVC components lend them-selves to reuse, reducing the cost and time required to develop future applications.

Simplified Modifications   Because they remain largely independent, programmers can often change one MVC component without affecting the others.

Concurrent Development   Functional segregation across modules allows multiple developers to work in parallel.

Component Distribution   Software architects can position (deploy) application modules anywhere on the globe. They can also replicate components and locate them in multiple data centers. Thus, MVC-based applications are typically robust in the face of component, network, and site failures.

However, like any design approach, the MVC architecture is not a panacea and has some inherent shortcomings, including:

Increased Complexity   As we create more modules, we increase the number of "moving parts" in an application. Tracking them can become problematic.

Network Dependence Component distribution relies on network speed, reliability, and availability. A distributed MVC-based application will only perform as well as its underlying communication infrastructure.

Security Concerns As you add components and network nodes to a system, you increase its points of vulnerability.

Despite these shortcomings, MVC's design advantages make it the architecture of choice for most Web-based applications.

ADVANCED TOPIC: SERVICE-ORIENTED ARCHITECTURE

The previous section demonstrated how MVC *Controllers* allow multiple *Views* concurrent access to the identical *Model* (data set). But is there any inherent reason why *Views* must limit their interaction to only one *Controller*? In other words, what prevents a *View* from accessing data from multiple *Controllers* and aggregating it on the user's screen? Well, nothing!

Consider the diagram presented in Figure 10.5; it depicts the architecture of a typical Web-based shopping application. Based on this design, the browser (i.e., the *View*) can interact with one or more of the *Controllers* to complete each request.

For example, while a user decides which items to buy, the browser repeatedly requests data from the Inventory System to display product information and images on the screen. When the user elects to purchase something, the browser interacts with the Inventory System to add the item to the Shopping Cart. When the user checks out, the browser uses both the Inventory System and the Shopping Cart to record purchases in the user's Account.

The point is that the three systems (Inventory, Shopping, and Account) operate independently and need not know of the others' existence. It's only the application code running in the browser (i.e., the *View*) that integrates them. We call this a *Point-in-Time Application* or PITA.

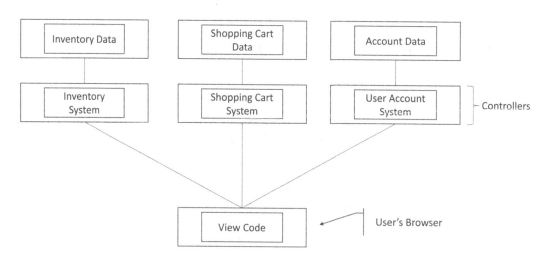

FIGURE 10.5 Point-in-Time application (PITA).

A *Point-in-Time Application* (*PITA*) integrates multiple components to satisfy a specific user request at a given moment in time.

A PITA integrates all components required to satisfy a specific user request at a particular instant. The next user request might require assistance from a different combination of *Controllers*.

To best leverage such a design, it would be advantageous if every *Controller* made its functionality available in small "chunks," allowing *Views* to "pick and choose" what they need as if ordering *à la carte* from a restaurant menu. Namely, if all *Controllers* offered a suite of individual "services," *Views* could combine and integrate them selectively.

This extension to the MVC architecture is so prevalent that it has its own name: *Service-Oriented Architecture* or SOA.

A *Service-oriented Architecture* (*SOA*) is a design model wherein individual *services* are available via network connections to all authorized application components.

Under SOA, *Controllers* provide *services* to authorized *Views* or other *Controllers* and components. We define a *service* as follows.

A *Service* is a unit of independent, self-contained functionality that performs one task, and its implementation is *opaque* (i.e., anyone using it doesn't have to know how it works).

Each SOA *service* has the following properties:

- It represents a single logical action that has specific, defined outcomes
- It's discrete and doesn't rely on any other *service*[15]
- It's opaque; callers need not be aware of its design or implementation

Extending the concept, software designers create *service suites* that implement a complete solution; we call this an *abstraction*.[16]

For example, a "shopping cart" *abstraction* might include *services* such as:

- Create a shopping cart
- Add an item to the cart
- Delete an item from the cart

---

[15] Note that there is the concept of *composite services* that may call one or more other services.
[16] Yet another type of *abstraction*.

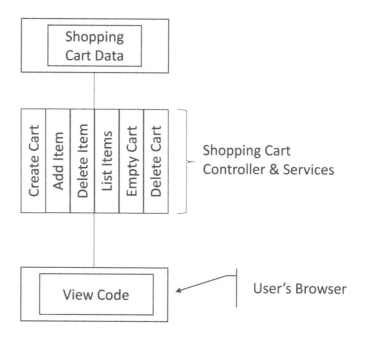

FIGURE 10.6   Service-Oriented Architecture (SOA) services.

- Empty a shopping cart

- List the items in the cart

- Delete shopping Cart

Because service suites often comprise individual *Controllers*, we often represent them as depicted in Figure 10.6.

SUMMARY

In this chapter, we discussed what constitutes a computer application. We also defined software architecture and discussed two common types: MVC and Service-Oriented. We also discovered how SOA could form the basis of Point-in-Time applications.

In Chapter 11, we'll focus our attention on the techniques IT professionals use to build software systems.

Software

How It's Built

Software is a great combination between artistry and engineering.

BILL GATES

INTRODUCTION

Thus far, we've discussed the distinction between hardware and software, how digital components communicate, and how instructions execute inside the CPU. In this chapter, we'll introduce the remaining puzzle pieces and learn how developers build software.

Software development is as much art as it is science. Indeed, if you assigned a hundred professional programmers the identical task, you'd receive a hundred unique solutions in return. They might all function correctly, but they would not look alike.

At first blush, this might seem surprising—even alarming—but it's no different than any other human undertaking requiring some degree of creativity and interpretation. For example, consider asking every meeting attendee to write a set of minutes—no two would read identically.

Although human ingenuity plays a significant role in software development, we nonetheless can't allow programmers to "run amok." Thus, through the adoption of formalized development methodologies, the IT industry has defined and organized the tasks required to design and build professional-caliber software applications. That is the focus of this chapter.

WHAT IS COMPUTER PROGRAMMING?

The purpose of computer programming is to develop a specific sequence of instructions that automates a given task. More formally:

Computer Programming is a set of tasks and procedures conducted by software developers that, when completed, generate a set of executable instructions that achieve a specific result.

However, software development doesn't happen in a vacuum. Professional-caliber programming requires that developers have expertise in both computer science and the subject matter they're automating. Consider developing an accounting program. The programmers must understand concepts like debit, credit, and depreciation to write the necessary code. We call this *domain-specific knowledge*.

In addition to domain-specific expertise, professional programmers must also understand the tools and techniques of software development. Serving as a foundation for this knowledge is a thorough understanding of one or more programming languages.

WHAT IS A PROGRAMMING LANGUAGE?

Let's begin our discussion of programming languages by comparing them with natural languages. The objective of both is communication: expressing ideas, sharing information, and issuing instructions. They also rely on standardized rules of construction consisting of a defined vocabulary, formalized grammar, and permissible operations.

However, that's where the similarities end. Natural languages are typically large, riddled with ambiguity, and evolve, well, naturally. Whereas computer languages are artificial constructs with a finite set of rules that change slowly.

To understand the differences, consider the following English sentence:

I didn't start my car because my friend wasn't in it.

Although it's grammatically correct, there are two valid interpretations of that sentence:

1. I didn't start my car because I was waiting until my friend got in.

2. I didn't start my car for a reason that had nothing to do with my friend not being in the car.

Another classic example shows us how the use of punctuation can change the meaning of an English sentence:

Let's eat, Grandma.

and,

Let's eat Grandma.

In the first case, we're urging Grandma to serve the food; in the second, Grandma is on the menu.

Humans can work through such vagueness—especially when conversing. However, ambiguity of this sort would wreak havoc in the IT world: each *expression* or *statement* in a computer language—the equivalent of a sentence in a natural language—must have one and only one interpretation. Otherwise, the same statement could yield a different result with every execution. That would be anathema in a world that expects accurate, repeatable results.

Consequently, computer languages are small (comparatively speaking) and impose strict usage and syntactical rules. Thus, we can define a programming language as follows:

> A *Programming Language* is a formalized set of general-purpose instructions that produce a precise result when combined and executed in a specific arrangement.

Computer language design is complicated—this is a field of study in itself. Language creators must consider many attributes: What sort of problems will it solve? What types of constructs will it provide? What environments will it support?

However, the most important design aspect is consistency. Regardless of how developers use and combine a programming language's constructs, every statement must have one and only one interpretation. As noted above, there can be no ambiguity.

However, before we start writing code, let's review the evolution of computer languages.

A Brief History of Programming Languages

Although most of us cannot remember life without them, computers have not been around all that long. As noted in Chapter 2, we can make a case that computing has been in existence—albeit in a primitive form by today's standards—as early as the 1800s. However, from a practical perspective, the types of computers we are familiar with date back to the early 1950s.

Nonetheless, despite the industry's relative infancy, thousands of computer languages have been developed. Some are general-purpose; others address specific needs. What follows is a brief history of some of the more significant developments in computer language evolution.

1800s   Ada Lovelace developed a program to compute Bernoulli numbers. Historians consider this the first computer program and Ada Lovelace the first computer programmer.

1949   The first Assembly Language.

1957   J. Backus developed the Fortran (FORmula TRANslation) language. Despite its early origins, it's still in use today.

1958   A committee of American and European computer scientists developed a language called Algol (Algorithmic Language). Although it never achieved independent popularity, Algol served as the foundation for many other influential computer languages.

1959   Rear Admiral Dr. Grace Murray Hopper developed a general-purpose computer programming language called COBOL (COmmon Business-Oriented Language). COBOL was significant because it was one of the first languages that could execute on many different brands of computers. COBOL is still in use today.

1959   John McCarthy of MIT developed LISP (LISt Processor), the first language used for Artificial Intelligence (AI) research. Among many innovative ideas, the language allows programs to modify themselves.

1964   While working at Dartmouth College, John G. Kemeny and Thomas E. Kurtz developed BASIC (Beginner's All-purpose Symbolic Instruction Code). Created as a pedagogical tool, BASIC is often the first language taught to beginning computer science students.

1970   Niklaus Wirth developed a language called Pascal. (He named it in honor of the famous French mathematician Blaise Pascal.) Although primarily created as an instructional tool for computer science students, it also served as the primary programming language in early generations of Apple computers.

1972   While working at Bell Labs, Dennis Ritchie created The C Programming Language. (It extended a language called, oddly enough, *B*.) Ritchie developed C to support the design and implementation of the UNIX Operating System. It also influenced many other computer languages such as Java, Python, Perl, and C#. It's still widely used today.

1972   While working at IBM, Donald Chamberlin and Raymond Boyce developed a language called Structured Query Language (SQL, pronounced "sequel"). Based on Relational Algebra (developed by Dr. Edgar Codd while also working at IBM), SQL supports the querying and manipulation of data stored in a relational database. Variants of SQL are in widespread use today.

1985   While working at Bell Laboratories, Bjarne Stroustrup published the first edition of his book, *The C++ Programing Language*. An extension of C, C++ is an object-oriented (OO) language that is still in use today.

1991   While working as a CERN[1] contractor, Tim Berners-Lee published a document called "Hypertext Markup Language (HTML) Tags", which proposed a language that could format text and images on Web pages. As a result of his innovation, historians consider Mr. Berners-Lee the inventor of the World Wide Web.

1991   Guido Van Rossum developed Python (named as an *homage* to the British comedy group *Monty Python*). Widely used, Python serves the needs of many development organizations.

---

[1] The European Organization for Nuclear Research.

1995   While working at Sun Microsystems, James Gosling originally developed Java to support small devices such as cable
boxes and hand-held computers. After subsequent improvements enhanced its power, Java became the *lingua franca* of
the Web. To say Java is everywhere is an understatement.

1995   Brendan Eich designed JavaScript to enhance browser
development. Today, almost every website you visit uses
JavaScript.

2000   Microsoft developed C#. Based on C++ and Java, it's used
primarily in Microsoft components.

This list is but a tiny sample. Existing languages continue to evolve, and the IT industry continues creating new ones to address emerging needs.

Classes of Computer Languages

One reason why there are so many computer languages is that they each possess inherent characteristics that address a particular need. Such design diversity is not unique to Computer Science. For example, in the world of carpentry, there are dozens of distinct types of hammers—each intended to satisfy a specific task.[2] Similarly, when writing large-scale computer systems, developers will often use a different language to implement each component: *the right tool for the right job*.

The sections that follow discuss several classes of programming languages. However, please note that some of them fit into multiple categories. For example, C++ is both an *object-oriented* and a *compiled* language. In contrast, Java is both an *object-oriented* and an *interpreted* language. (We'll discuss what these terms mean in the following sections.)

Machine Languages

The most fundamental class of computer language is the CPU's instruction set. As we've seen in earlier chapters, every machine language *opcode* directs the CPU to execute one primitive instruction. Although every executable program ultimately runs as a series of opcodes (we'll describe how this happens later in the chapter), programming directly in machine language has many shortcomings:

- It's tedious and error-prone: Developers must use 1s and 0s to write programs.

- Programs written in machine language are challenging to read; thus, making changes to them is often problematic.[3]

- Machine language instructions are CPU-specific: Programs written using one CPU's instruction set cannot run on other processors.[4]

---

[2] That said, I find a butter knife to be the tool of choice for most household repairs. You know what "they" say, when all you have is a butter knife, everything looks like a … well, I'm sure you know what I mean.

[3] We refer to this as *readability*.

[4] We refer to this as *portability*.

Because programming in a machine language is so tedious, the IT community developed computer languages with increasingly greater degrees of *abstraction*. As we will see, when we increase the levels of abstraction, we reduce the levels of complexity for programmers.

Assembly Languages

The first level of abstraction is *assembly language* (or *assembler language* or just *assembler*). Although still low-level, assembler languages provide human-readable mnemonics for every machine-language instruction. Thus, instead of typing a string of 1s and 0s as in:

```
01101100 000111001 0110001
```

Programmers would instead code something like the following:

```
MOV B, A
```

As I'm sure you'd agree, assembler mnemonics are easier to read and write. This simplicity is especially beneficial when modifying existing programs—the easier code is to read, the easier it is to comprehend.

To prepare an assembler program for execution, developers use a tool called, oddly enough, an *assembler* to convert mnemonics into opcodes. Returning to the example above, if a program contained the statement:

```
MOV B, A
```

the assembler would generate the following output:

```
01101100 000111001 0110001
```

Though easier than coding machine instructions, assembler language programming suffers several deficiencies:

- Writing assembler code is also tedious and highly error-prone.

- It also lacks portability because of the one-to-one relationship between assembler mnemonics and machine language opcodes.

As we will see, to address these shortcomings, the computer science community developed *high-level languages* that increase the levels of abstraction.

Imperative vs. Declarative

The first way to classify high-level computer languages is by whether they are *imperative* or *declarative*.[5] When using *imperative* languages, programmers specify the operations—i.e., the sequence of instructions—that, when executed, produce the desired result.

---

[5] As compared to natural languages, these terms differ slightly when used to describe computer languages.

For example, consider the following C++ program that calculates a rectangle's area based on its length and width.[6]

LISTING 11.1   EXAMPLE C++ PROGRAM

```
#INCLUDE <IOSTREAM>
USING NAMESPACE STD;
INT MAIN ( INT AC, CHAR* AV[] )
{
        INT     AREA, LENGTH, WIDTH;

        WIDTH = 5;
        LENGTH = 10;
        AREA = LENGTH * WIDTH;

        COUT << "THE AREA IS: " << AREA << ENDL;
}
```

When it executes, the program will display the following output:

LISTING 11.2   SAMPLE PROGRAM OUTPUT

```
THE AREA IS: 50
```

Don't worry if you don't understand every statement. (We'll review other programming examples in more detail later in the text.) Just keep in mind that the programmer (me in this case) specified the length (LENGTH = 10;) and width (WIDTH = 5;) of the rectangle, then computed its area (AREA = LENGTH * WIDTH;).

The last statement displays the result:

```
COUT << "THE AREA IS: " << AREA >> ENDL;
```

In contrast, when using a *declarative* language, programmers specify the desired result, not the operations required to produce it. To demonstrate this approach, let's assume we had a database that contained a list of employees, as depicted in Table 11.1.

TABLE 11.1   Employee Table

Employee#	First_Name	Last_Name	Start_Date	Hourly_Rate
111	Jane	Smith	1/1/20	25.00
222	John	Doe	2/1/20	24.00
333	Sal	Minella	3/1/20	23.00
444	Alice	Wonderland	4/1/20	22.00

---

[6] The intent of this listing is illustrative and pedagogical; it's NOT an example of a well-crafted professional-caliber C++ program.

If asked to create a report that displayed the Employee Number, Last Name, and Hourly Rate for each worker, we could use the following SQL statement:

LISTING 11.3   EXAMPLE SQL STATEMENT

```
SELECT Employee#, Last_Name, Hourly_Rate FROM Employee_Table;
```

When executed, the output of this statement would look something like this:

LISTING 11.4   SAMPLE SQL PROGRAM OUTPUT

```
111 SMITH 25.00
222 DOE 24.00
333 MINELLA 23.00
444 WONDERLAND 22.00
```

Again, don't worry about the syntax. Instead, please focus on the crucial point: the SQL statement specified *what* we wanted to display, not *how* to do it. The "how" is part of the language's semantics.

Structured Programming Languages

Structured programming languages represent the next level of abstraction. Developed in the mid-to-late 1960s, structured languages improve the power, readability, and clarity of computer programs. Some popular examples include C, ALGOL, and PASCAL.

Although the syntax and semantics of structured languages may vary, they all share the basic control constructs described below. Developers may freely combine them to achieve a desired result.

Sequence   The ordered execution of one statement after another.

Selection   Choosing which instructions execute based on one or more conditions. For example, most web pages allow users to either sign-in or create a new account. The code to implement that type of processing might look something like this:

```
IF( USER PRESSED THE "SIGN-IN BUTTON")
THEN
    EXECUTE THE "SIGN IN" CODE
ELSE
    EXECUTE THE "NEW ACCOUNT" CODE
END
```

Based on the result of the *conditional expression* (i.e., Did the user press the sign-in button?), the program will either execute the sign-in code or the new account code.

Note that I didn't use an actual programming language in this example. Instead, I expressed the logic using a construct called *pseudocode.*

> *Pseudocode* allows developers to describe programming logic using human-readable constructs without becoming mired in a specific language's syntactic details.

Like an artist's sketchbook, pseudocode allows programmers to capture and convey programmatic ideas in a convenient, human-readable format.

Iteration Repeat a sequence of instructions until some condition causes the looping to terminate. For example, the pseudocode to add the numbers 1 through 10 (inclusive) might look something like this:

```
SUM = 0                      // INITIALIZE SUM TO 0
COUNTER = 1                  // START COUNTER AT 1

// LOOP UNTIL COUNTER REACHES THE VALUE 11
WHILE( COUNTER IS LESS THAN 11 )
DO
    SUM = SUM + COUNTER       // ACCUMULATE THE SUM
    COUNTER = COUNTER + 1     // INCREMENT THE COUNTER
DONE
```

Execution begins with the initialization of the variables, SUM and COUNTER, to 0 and 1, respectively. The WHILE loop's body (i.e., the code positioned between the DO and DONE syntactical placeholders) repeatedly executes until the variable, COUNTER, reaches the value 11. During each loop iteration, the code aggregates the sum (SUM = SUM + COUNTER) and increments the counter (COUNTER = COUNTER + 1).

Note that I've introduced another new statement type in this example. By convention (and language specification), all text following double forward slashes ("//") are *comments* written by programmers for other programmers. They do not execute. When algorithms become complex, it's often difficult to understand the meaning of code written by another programmer. Developers include comments to help clarify programming logic.

Blocks Programming *blocks* allow developers to group multiple statements together and treat them as a single unit. As an example, consider the pseudocode in the iteration section above. Collectively, all statements within the body of the WHILE loop form a programming block and execute during each iteration.

Developers may also assign names to programming blocks.[7] Using this construct, developers can reuse a block by referencing its name rather than reinserting the identical code each time they need to use it.[8]

As an example, I converted the code from the iteration section above into a named block (formally, we refer to this as a *function*). To aid readability, I've highlighted the key changes/additions in gray and added further comments.

```
SUM_INTERGERS( MAX_VALUE )    // FUNCTION DECLARATION
BEGIN                         // BEGIN THE BLOCK
  SUM = 0                     // INITIALIZE SUM TO 0
  COUNTER = 1                 // START COUNTER AT 1

  // LOOP UNTIL COUNTER REACHES MAX_VALUE
  WHILE( COUNTER IS LESS THAN MAX_VALUE+1 )
  DO                          // BEGIN BODY OF LOOP
    SUM = SUM + COUNTER       // ACCUMULATE THE SUM
    COUNTER = COUNTER + 1     // INCREMENT THE COUNTER
  DONE                        // END BODY OF LOOP
  RETURN( SUM )               // RETURN THE RESULT
END                           // END THE BLOCK
```

As part of its declaration (the first line), I named the function SUM_INTEGERS. The BEGIN and END keywords delineate its boundaries (i.e., its *body*).

The declaration also specifies that SUM_INTEGERS accepts one *parameter*, which I've named MAX_VALUE. Parameters are placeholders for values that functions may receive when invoked. Thus, instead of stipulating the value 11[9] as in the iteration example above, MAX_VALUE can contain a different number each time we call the function.

---

[7] In computer science, there are several types of named code blocks: *subroutines, functions, procedures,* and *methods.*

[8] Called *code reuse*, this is a *best practice* in software development.

[9] In computer science we would call this *hard coding* values. It's not a best practice and is strictly *verboten.*

Given this function declaration, we can sum the integers from 1 to 10 as follows:

```
ANSWER = SUM_INTERGERS( 10 )
```

After this statement executes, ANSWER contains the value 55.[10]

However, because of the function's design, we can pass it any valid integer value. For example, to sum the numbers from 1 to 100, we would invoke SUM_INTERGERS as follows:

```
ANSWER = SUM_INTERGERS( 100 )
```

This expression assigns ANSWER the value 5,050.[11]

Although it requires additional effort to package functionality into named blocks, this is a common technique programmers use to avoid writing duplicate code.

Object-Oriented Programming Languages

While they incorporate many of the same basic constructs of structured languages (i.e., sequence, selection, iteration, and named blocks), OO languages alter the way developers approach software design. In structured languages, programmers tend to think linearly, creating blocks of code that execute sequentially. In contrast, when using OO languages, programmers think in terms of *objects* and their *interactions*. This paradigm is much more consistent with the way we view the world.

To highlight the difference between the two approaches, consider a typical banking transaction: transferring money from a savings account into a checking account. When you thought about that just now, I'd be willing to bet that you didn't think in terms of *code blocks*. More likely, what came to mind were *objects* like *bank* and *account*. In addition, you envisioned the transfer as an operation (or *behavior*) involving the two account objects: the savings account object allows withdrawals; the checking account object permits deposits.

Similarly, when coding in OO languages, developers design programs that closely align with real-world concepts: they identify significant *objects* and ascribe *behaviors* to them. This is a powerful design paradigm that, when implemented correctly,[12] engenders well-constructed programs that are (relatively speaking) simple to build and comprehend.

To demonstrate the power of this model, let's look at a simple OO design for our banking example. As you read through the code below, please keep in mind that in OO languages, we encapsulate *behaviors* in *methods* (aka, *functions*), which are another type of named code block.

---

[10] In binary, of course!

[11] Again, in binary.

[12] On the other hand, there are few things in life as convoluted and confusing as a poorly designed object-oriented program.

Let's begin with the SAVINGS_ACCOUNT object.

As depicted in the pseudocode in Listing 11.5,[13] *objects* contain two main sections: DATA and METHODS. In the DATA section, we define variables (i.e., "lockers") such as CURRENT_BALANCE and ACCOUNT_NUMBER.

The METHODS section contains the *behaviors* associated with the object. The SAVINGS_ ACCOUNT object has two: DEPOSIT() and WITHDRAWAL().

LISTING 11.5   SAVINGS ACCOUNT OBJECT

```
OBJECT SAVINGS_ACCOUNT
BEGIN
        DATA                    // VARIABLES
            FLOAT     CURRENT_BALANCE
            INTEGER ACCOUNT_NUMBER

        METHODS                 // BEHAVIORS
            DEPOSIT( AMOUNT )
            BEGIN
                    // CODE FOR THE DEPOSIT METHOD WOULD APPEAR HERE
            END
            WITHDRAWAL( AMOUNT )
            BEGIN
                    // CODE FOR THE WITHDRAWAL METHOD WOULD APPEAR HERE
            END
END
```

Except for its name, the description of the CHECKING_ACCOUNT object (Listing 11.6) is remarkably similar.[14]

LISTING 11.6   CHECKING ACCOUNT OBJECT

```
OBJECT CHECKING_ACCOUNT
BEGIN
        DATA                    // VARIABLES
            FLOAT     CURRENT_BALANCE
            INTEGER ACCOUNT_NUMBER

        METHODS                 // BEHAVIORS
            DEPOSIT( AMOUNT )
            BEGIN
                    // CODE FOR THE DEPOSIT METHOD WOULD APPEAR HERE
            END
            WITHDRAW( AMOUNT )
            BEGIN
                    // CODE FOR THE WITHDRAWAL METHOD WOULD APPEAR HERE
            END
END
```

---

[13] For those of you scoring at home, we are ignoring the concept of CLASS during this discussion.

[14] OO languages allow objects like CHECKING_ACCOUNT and SAVINGS_ACCOUNT to share code for methods like DEPOSIT and WITHDRAWAL. This construct, called *inheritance*, is another type of code reuse but, alas, is beyond the scope of this book.

Note that, for both the SAVINGS_ACCOUNT and CHECKING_ACCOUNT objects, I didn't include any pseudocode for their DEPOSIT() and WITHDRAWAL() methods. For our purposes, the details of their implementation are irrelevant.

However, I do want to highlight an important point: objects manage both data and behaviors *privately*.[15] Only the developers who wrote the code for each object need to understand their implementation details. Programmers who use the objects need only be aware of the behaviors they provide; they don't need to know how they work.

Given these two objects, another developer might write a *transfer* method as depicted in Listing 11.7.

LISTING 11.7   TRANSFER FUNCTION

```
TRANSFER_FROM_SAV_TO_CKING ( CHK_OBJECT, SAV_OBJECT, AMOUNT )
BEGIN
            SAV_OBJECT.WITHDRAW ( AMOUNT )
            CHK_OBJECT.DEPOSIT ( AMOUNT )
END
```

The TRANSFER_FROM_SAV_TO_CKING() function accepts three parameters: the checking account object (CHK_OBJECT), the savings account object (SAV_OBJECT), and the dollar amount of the transfer (AMOUNT). When called, the function invokes the WITHDRAWAL() method of SAV_OBJECT and the DEPOSIT() method of CHK_OBJECT. (In many OO languages, the dot "." notation is the syntax used to select and invoke an object's methods.)

We can execute a $100.00 transfer as follows:

```
TRANSFER_FROM_SAV_TO_CKING ( MY_CHK_ACCOUNT, MY_SAV_ACCOUNT, 100.00 )
```

Miscellaneous Language Categories

There are many other classes of programming languages that are beyond the scope of this text. Nonetheless, we've included a list below with some examples for those readers who wish to learn more about them.

- Functional Programming Languages: SML, Scala, Erlang

- Logic-based Programming Languages: PROLOG

- Command-line Languages: Windows PowerShell, BASH

- Scripting Languages: AWK, Perl

- Web Languages: HTML, JavaScript

- Markup Languages: HTML, XML

---

[15] In computer science, we refer to the technique of restricting access to implementation details as *encapsulation*.

HOW DO PROGRAMS EXECUTE?

After we finish coding a computer program, it doesn't just run all by itself. On the contrary, we need to complete a specific set of tasks to prepare the source code for execution.

The sections that follow describe the two most common execution paradigms: compilation and interpretation.

Compilation

In prior discussions, we saw examples of programs written in machine language and commented on how tedious it is to develop solutions using only the CPU's primitive instruction set. We also noted how this prompted the development of high-level, human-readable languages.

Unfortunately, computers can only execute machine-language instructions. Consequently, to make programs developed in high-level languages executable, we need to *translate* them into a form that allows them to run. We refer to this process as *compilation*.

> *Compilation* is the process of translating a computer program written in one language (called the *source language*) into another language (called the *target language*).

Typically, the source is a high-level, human-readable computer language, and the target is machine code. However, this is not always the case. For example, some compilers can convert COBOL into Java.[16]

Compilation is comparable to translating into English a novel written in French. There is, however, an important distinction. Because they are so complex, quirky interpretations often occur with natural language translations. (I'm sure everyone has heard the expression, "Something got lost in translation.") For example, we cannot translate the idiom "A chip off the old block" *literally*. We must convert such expressions into phrases that are *equivalent* in meaning in the target language.

In contrast, there is no "wiggle room" with computer language compilation—semantic precision counts. That is, compiled programs must execute exactly as specified by the statements in the source language.

Let's examine this process more closely. As depicted in Figure 11.1, the compiler reads code contained in a *source file*. When it completes, it stores the machine code in an *executable file* that can run on the local machine.

FIGURE 11.1   Compilations process.

---

[16] For obvious reasons, such programs are often called *translators*.

Time for an example. Listing 11.1 contains the source code for one of the simplest programs you can write in the C Programming language.[17] It's called *Hello World!*.[18]

LISTING 11.8   HELLO WORLD!

```
#INCLUDE <STDIO.H>                    // IGNORE THIS FOR NOW
MAIN ()
{
        PRINTF ( "HELLO, WORLD!\N" );    // SAY HELLO TO THE WORLD!
}
```

When this program executes, it displays the text, HELLO WORLD!, on the screen. Not all that exciting, I know. But it does serve our needs.

To prepare it for execution, I created a file called HELLO.C[19] and typed in all the code contained in the listing. After I saved and closed the file, I executed the following command:

G++ HELLO.C

The program, G++, is the C compiler's name on the system I used to create this example.[20] When it finished running, the compiler produced a file called A.EXE, which is the default name used by G++ for every executable file it creates.[21]

I then executed the command, A.EXE, which displayed the following text on my screen:

HELLO, WORLD!

Despite its simplicity, this example does highlight some of the tasks undertaken by professional developers to prepare programs for execution. We'll expand on this knowledge later in the chapter when we discuss the software development lifecycle.

Interpretation

Another common way that programs execute is through a process called *interpretation*. With this approach, we don't compile a program into machine language. Instead, we translate the source code into an intermediate representation, usually called *bytecode* (or *p-code*[22]), and then another program, called an *interpreter*, executes the intermediate code on our behalf.

An *interpreter* is a computer program that directly executes instructions without requiring compilation.

---

[17] In any language for that matter.

[18] Dating back to the early 1970s, Brian Kernighan holds the distinction of creating the first version of the *Hello World* program.

[19] To do this, I used an editor called NOTEPAD++. It's like Microsoft Word for software developers.

[20] I used Cygwin running atop MS Windows. Feel free to ignore the weird name of the compiler.

[21] You can instruct G++ to change the name of the executable file.

[22] Shorthand for *portable code*.

Though accurate, that definition may seem somewhat ambiguous. Let's see if we can bring some clarity to this process.

As I've repeatedly stated throughout this text, one of the most fundamental and powerful design tools in computer science is *abstraction*. Thus far, we've seen many examples of this technique: machine language instructions represented as opcodes, the DARPA Layered Network Model, the file/folder paradigm for disk storage, the desktop metaphor for user interfaces, etc.

Extending that idea, what's to stop us from creating an abstraction of a *computer*? That is, why can't we write software that simulates hardware? The answer is: nothing! We call such a program a *Virtual Machine* (or VM).

> A *Virtual Machine* is a software representation of a computer system that emulates the underlying hardware functionality.

As you may recall from Chapter 4, computers contain many components, ALUs, CPUs, registers, buses, memory, etc. VMs simulate all those components in software.

Time for another example. To see interpretation in action, let's implement our *Hello World* program in one of the most widely used interpreted languages: Java.[23]

Listing 11.6 contains the source code for a file I created called HELLOWORLD.JAVA.

LISTING 11.9   HELLO WORLD IN JAVA

```
// Hello World in Java
class HelloWorld
{
        public static void main( String[] args )
        {
                System.out.println( "Hello, World!" );
        }
}
```

Again, please don't get mired in language constructs. Like its C Language counterpart, this program displays the text, HELLO WORLD!, on the screen when executed.

To compile this program, I ran the following command:

```
javac HelloWorld.java
```

The program, JAVAC, is the name of the Java compiler on my system. However, unlike the C compiler (G++), JAVAC doesn't generate an *executable* file. Instead, it creates a file named HELLOWORLD.CLASS, which contains the intermediate *bytecode* produced by the translation process (see Listing 11.10).

---

[23] For the sake of thoroughness, I should note that there are also compiled versions of Java.

LISTING 11.10 BYTECODE FOR HELLO WORLD.CLASS

```
CLASS HelloWorld {
    HelloWorld();
        Code:
            0: ALOAD_0
            1: INVOKESPECIAL    #1
            4: RETURN

    PUBLIC STATIC VOID MAIN(JAVA.LANG.STRING[]);
        Code:
            0: GETSTATIC      #7
            3: LDC            #13
            5: INVOKEVIRTUAL  #15
            8: RETURN
}
```

We will not discuss how to decipher the bytecode representation; the important point is that it is not machine code and cannot run on any computer in its current form.

To execute the program, we run the following command:

JAVA HelloWorld

The program, JAVA, is the name of the Java Virtual Machine. When it interprets our program, it displays the following text on the screen:

HELLO, WORLD!

Please take a moment to note the differences between the Java example and its C Language counterpart. In this case, we didn't run the HelloWorld program directly as we did when we ran A.EXE. Instead, we invoked an interpreter (JAVA) to execute the instructions (*bytecodes*) contained in the file HelloWorld.CLASS on our behalf.

Comparison of Execution Techniques

Some languages support both compilation and interpretation. In such cases, you should expect identical results regardless of how you execute a given program.[24] (Unfortunately, that's not always the case.)

That said, the two execution options—*compilation* and *interpretation*—have relative advantages and disadvantages; I've summarized some of them below.

- As a rule, compiled code runs faster. The computer executes compiled instructions directly. Whereas with interpretation, there is an additional layer of software (i.e., the VM) required to execute the instructions.

---

[24] Bugs aside.

- Code executed via interpretation is more portable. With compiled languages, we must repeatedly recompile the source code for every target execution environment. In contrast, interpreted bytecode can run on any platform that hosts a VM.[25]

- When compiling code, you must fix errors before the program may execute. That is, the compilation process terminates if the compiler uncovers errors in your source code.[26] With many interpreted languages, you may not discover some types of semantical bugs until the program runs.

THE SOFTWARE DEVELOPMENT LIFECYCLE

For many disciplines, formal processes exist that ensure accurate and repeatable results. Consider the construction industry. Well-established project management techniques coordinate the delivery of materials and the scheduling of the various trades. There is also little latitude regarding what to build and how to build it: formal plans—*blueprints*—precisely specify every task tradespeople need to complete.

Unfortunately, that's not always the case when building computer systems. Developing comprehensive, professional-caliber software applications isn't as straightforward as is the case with other engineering domains.

First, unlike most other industries, every software application—indeed, every line of code—is built for the *first time*. Consider that a construction company may erect scores of similar houses in cookie-cutter fashion when developing a large residential housing project. However, even though I've designed and built over twenty order management systems in my career, no two were identical—each was uniquely designed to address a particular client's needs.

In addition, large software development projects may require coordinating the efforts of scores—if not hundreds—of professionals working in various capacities. Yet, if even a single team member loses focus, the quality of the product could suffer. Moreover, because writing software is both an art and a science, it leaves room for personal ingenuity (which can be extremely beneficial), individual shortcomings (which can be extremely detrimental), and miscommunication (which can wreak havoc).

Nevertheless, software development is not the Wild West of the engineering world. On the contrary, processes and guidelines exist that help ensure accurate results.

For the remainder of this section, we'll discuss the tasks required to build software applications. Collectively, these activities comprise the *Software Development Life Cycle* (SDLC for short).

> The *Software Development Lifecycle* is the management framework required to plan, design, develop, and deploy computer systems.

---

[25] In fact, the Java mantra is "Write once, run everywhere." More seasoned and cynical IT professionals counter that with, "Write once, *fix* everywhere."

[26] This refers to syntactical and semantical errors. Unfortunately, most logic errors can only be uncovered during testing (and use).

While reading through the material below, please keep in mind that, depending on such constraints as budget, schedule, formality, objectives, etc., project managers may elect to alter the chosen SDLC methodology or omit some of its tasks.

Requirements Specification

As I alluded to previously, we can't allow software developers to run amok. The first way project managers establish control is by informing the programmers what they must build—in precise detail. The best way to understand how this takes place is by example.

Let's assume we manage a warehouse and that to remain competitive, we need to automate more of its operations. Specifically, we want to design and build software programs that will assume control of day-to-day tasks.

During the initial planning sessions for this system, we consider the features we want to include in the new application. Should the system:

- Track inventory quantities
- Maintain product locations
- Direct robots to pick and pack orders
- All the above

We can't leave such decisions to our programmers' discretion—they don't understand the business as well as we do. They can certainly contribute to the discussion, but it's up to us—the project managers—to identify and specify the company's needs.

But how do we do that?

That's where *System Requirements* come into play. They specify, in varying degrees of detail, the functions we want the application to provide. Formally, we aggregate system requirements in a document called a *Requirements Specification.*

> A *Requirements Specification* defines and delineates the precise set of functions and features that a given software system must contain.

The industry uses formal requirements for both "shrink-wrapped" (i.e., purchased) products (such as Microsoft Office), as well as custom-designed software (such as the warehouse management system above). In both cases, the developers don't *guess*. On the contrary, project managers specify—through requirements—what the programmers should build.

Generally speaking, there are two types of requirements: Functional and Non-functional.

Functional Requirements specify the features the application must provide. Returning to our inventory example from above, that might include product tracking, order management, custom screen displays, warning messages, and so on.

Despite its seemingly oxymoronic name, *Non-functional Requirements* specify such items as performance, availability, and reliability. For example, we might stipulate that the system must respond to each user request in less than one second and must be available between the hours of 7:00 am and 7:00 pm, five days a week.

In varying degrees, all functional and non-functional requirements affect application design. However, the most important consideration when specifying requirements is that they are *testable*. For example, consider the following non-functional requirement:

> *The program must run fast.*

It's impossible to determine if a program meets that specification: it's too vague and entirely subjective. One individual may conclude that the program runs fine; another might argue that it's too slow.

With that in mind, let's recast that requirement as follows:

> *The program must respond to all queries in less than one second.*

Now, when evaluating our application's performance, we can use a stopwatch to determine empirically whether it meets the specification.

Analysis Phase

Depending on the project and the underlying development methodology (see below), the Analysis Phase may run concurrently with the Requirements Phase. This effort's overall objective is for the project team to acquire a sufficient understanding of the system to develop the application's architecture and refine its requirements (if appropriate).

During this phase, the technical team will perform the following tasks:

Requirements Review	Developers review requirements to determine if any inconsistencies, ambiguities, or omissions exist in the system specification. As a rule, the earlier the team uncovers such issues, the easier (and cheaper) they are to fix. (More on this later.)
User Meetings	System designers meet with users to put form to the requirements. That is, developers want to understand—firsthand—the users' needs and expectations.[27]
Review Existing Systems	Developers review any existing systems and procedures—whether manual or automated—to understand the objectives of the new (or replacement) system.
	Once this phase completes, the project team transitions to the Design Phase.

---

[27] In many organizations, internal policies and politics prohibit programmers from meeting directly with users. In such cases, technical staff must work with Business Analysts who (ostensibly) represent the users and their needs. The upside of this approach is that Business Analysts are allegedly proficient at expressing user requirements in technical terms. The downside is that this approach rarely succeeds because Business Analysts often lack the necessary skills to accomplish this task. Unfortunately, Business Analysts don't always understand the business as well as the users they purport to represent, and they are often unfamiliar with the technology used to build the system. Thus, in practice, this added layer of indirection often leads to confusion and friction among users and development staff.

Design Phase

There are two main objectives in the Design Phase: a robust system architecture and detailed component designs. In the sections that follow, we'll present some of the tools and techniques software designers use to produce these deliverables.

System Architecture

In Chapter 10, we discussed the importance of system architecture. I can't overstate its significance: like a building's foundation, it serves as the underpinning for all development—current and future.

For many large-scale projects, architectural design begins with the preparation of either a High-level Design (HLD) document or a System Architecture (SA) specification.[28] In addition to explanatory content, these documents identify the major hardware and software components and contain diagrams illustrating their interaction.[29]

Time for some examples. Let's begin with the diagram in Figure 11.2, which depicts a minimalist system architecture. Albeit simple, it does illustrate a typical *three-tiered application* architecture.[30] (Please refer to the discussion in Chapter 10 for further detail.)

From a practical perspective, the diagram in Figure 11.2 is far too abstract to allow programmers to begin coding; they need more detail before they can put fingers to keyboards.

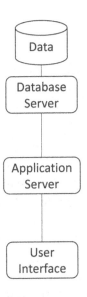

FIGURE 11.2   Sample system architecture diagram.

---

[28] These are two "flavors" of the same document. In IT, we often change the names of things to protect the innocent. Or maybe it's just a way to provide job security.

[29] These documents are the IT equivalent of blueprints.

[30] It may look like four, but because the database server also manages disk storage, we typically view the pair as a single component.

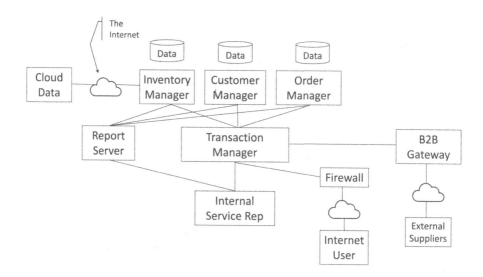

FIGURE 11.3   Detailed architecture diagram.

A more realistic example appears in Figure 11.3. The "cloud" icons symbolize Internet connections; the other boxes represent the components required to "stand up"[31] the system. For our purposes, we don't need to dive into the requirements of each element. Just note that such architectural diagrams define the system's extent and delineate the suite of modules comprising the solution.

State Diagrams

The logical architecture depicted above is only one *view* of a system. There are many other diagrams developers employ to capture various design elements of a complex solution. Some additional examples appear below; a professional design would include many others.

The first *view* we'll discuss is a *State Diagram*.[32] Application designers use these artifacts to specify system behavior in response to events. As a simple example, Figure 11.4 depicts a State Diagram for a lamp equipped with a pull chain to turn its light on and off.

Circles represent *states*. By convention, State Diagrams contain one *start state* and at least one *end state*. All the other circles represent *intermediate states* that the system can transition to/from in response to events. The text included in each circle defines *actions* (or processing) that the system must perform when entering[33] each state.

Arrows represent the list of permissible *events* to which the system must respond. They also indicate the *next state* to which the system will *transition* in response to that *event*.

---

[31] "Stand Up" is a term we use to represent the process of taking a system from conception to production. Watching that first transaction flow through a newly deployed application is one of the most satisfying moments in an IT professional's life. What can I say? We're easily amused.

[32] State Diagrams also come in many "flavors:" *Transition Diagrams, Transition Tables,* and *State Tables.*

[33] Although not shown in the figure, you may also specify actions that the system must perform when *exiting* a state.

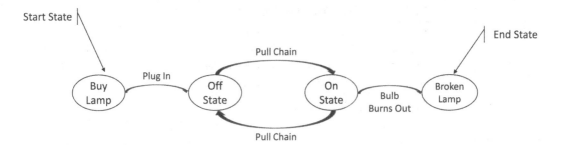

FIGURE 11.4   Example state diagram.

We interpret State Diagrams as follows:

- The system begins processing in the *start state*. In our example, this is the day we purchased the lamp.

- In every *state*, the system may receive *events*; arrows represent every *valid event* (for a given *state*) and indicate the *next state*.

- When an *event* occurs, the system executes the associated *actions* (i.e., the required processing) and then *transitions* to the *next state* (as indicated by the *arrow*).

- This processing continues until the system *transitions* to an *end state*.

Using the diagram in Figure 11.4, let's step through an example.

On the day we bring home our lamp, we can say that the system begins in the *start state*. When we plug in our new purchase, the system receives its first *event*; in response, it *transitions* to the "Off" *state*.

Note that there was only one valid *event* when the system was in the *start state*: "Plug-In." We can pull the chain all we want, but the bulb will not light because the lamp has no power. This is a fundamental feature of all state-based systems: they will only accept (i.e., respond to) *events* that are valid in their *current state*. They treat all other *events* as errors.

Once in the *Off state*, the lamp will respond to chain pull *events*, alternating between the *On* and *Off* states. Processing continues in this manner until the system *transitions* to the *Broken Lamp state* (*End state*) when the bulb burns out.

Please note the following:

- *States* allow programs to track their progress. When in a given state, a program may assume that all processing required to have arrived there has been completed successfully.

- Some *states* may permit multiple *events* (e.g., the "On" *state* in our example).

- Different *states* may accept the same *events* (e.g., the "Off" and "On" *states*).

- *States* may *transition* back to the same *state* (not shown in our example).

TABLE 11.2   State Transition Table

State	Event	Action	Next State
Buy Lamp	Plug-In Lamp	N/A	Off
Off	Pull Chain	Turn On Light	On
On	Pull Chain	Turn Off Light	Off
On	Bulb Burns Out	N/A	End

Transition Tables

As depicted in Table 11.2, we can represent State Diagrams in a tabular format. Known as *Transition Tables*, developers find this *view* extremely helpful because it clearly defines system functionality.

Because they are simple to draw, designers typically use State Diagrams to capture system state and event processing. Once completed, they often convert them to Transition Tables to simplify the efforts of programmers.

Data Architecture Diagrams

By far, the most valuable resource in any system is its data. Ultimately, regardless of what an application does, how powerful it is, or how slick it looks, the only real concern is how it manipulates its data. (In law enforcement, you follow the money; in software development, you follow the data.)

Thus, in modern software design, it's *data first*. As a result, an application is only as good as its data architecture.

System designers use several *views* to represent data architecture in large-scale applications. The first, called a *Data Flow Diagram* (or DFD), captures how data travels through a system. A simple example appears in Figure 11.5.

As the diagram indicates, when a Customer enters an order, the Order Management component sends data to the Warehouse module which, in turn, forward data to both the Shipping and Inventory subsystems. In this way, Dataflow Diagrams capture the flow of information through a system and highlight component interdependence.

Another way to view data is to understand the relationships among related elements. For example, when a customer completes a purchase, how does the system use order data to

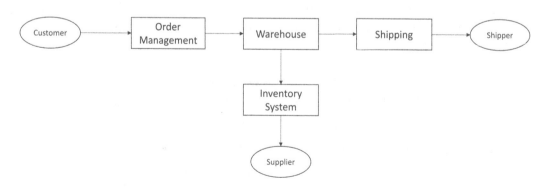

FIGURE 11.5   Data flow diagram.

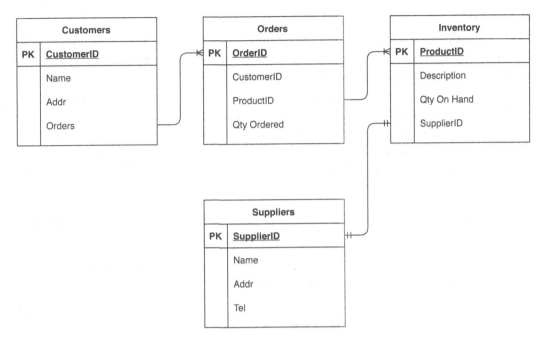

FIGURE 11.6   Entity-Relationship diagram.

ensure that warehouse personnel select and package the desired product? Or, when shipping orders, how can we ensure that the correct address appears on the address label?

We use a *view* called an Entity-Relationship Diagram (or ERD for short) to identify data elements (in collections called TABLES) and capture their interrelationships. The ERD depicted in Figure 11.6 contains four TABLES: CUSTOMERS, ORDERS, INVENTORY, and SUPPLIERS.

Note that the design uniquely identifies each customer order with a CUSTOMERID and each order with an ORDERID. As with CUSTOMERIDs, each ORDERID is unique and represents exactly one order in the ORDERS table. Similarly, orders contain PRODUCTIDs that represent individual products in the PRODUCT table.

Using this design approach, we can follow the "trail" from Customer to Product to Supplier without replicating or duplicating data. We can also identify the relationships among the various data elements.[34]

View Adoption

Software designers don't create every view for every system they build. Generating and maintaining diagram's requires time, and time is money. Thus, depending on a given system's complexity, designers will weigh each view's benefits and costs when deciding to use it. In line with that approach, some of the development methodologies we'll discuss later in the chapter require only a subset of these diagrams. Indeed, some forgo all of them.

---

[34] Although beyond this book's scope, we also use ERDs to create schemas for the databases that serve as the electronic custodians of application data.

Buy vs. Build

Despite the detail provided by all the diagrams discussed above, we're still not ready to begin coding. First, as described in the next section, each architectural component requires further refinement before we can "turn the programmers loose."

In addition, and more importantly, we don't necessarily have to build every module "from scratch." In most development projects, system designers will conduct "buy vs. build" reviews and, in many cases, opt to purchase components rather than develop them "in-house." For example, most IT shops would typically buy database systems rather than write their own.[35]

Some of the evaluation criteria that designers consider during the buy-vs.-build decision include:

- The benefits of custom-built software vs. acquiring general-purpose packages that may include features appealing to a broader audience.

- License fees for purchased products vs. the costs associated with development and maintenance of software developed "in-house."

- Can the cost of purchased products be offset by the savings realized by shortened development schedules?

After conducting this evaluation, designers prepare a list of components they will buy and identify those that the development team will build.

Program and Component Design

As discussed above, the objective of system architecture is the identification and interaction of major system components. Once defined, we must design each of these modules before programming can commence. Two of the key design attributes in this phase are *abstraction* and *encapsulation*.

As a brief reminder, *abstraction* is the distillation of the essential and relevant attributes of a problem domain to simplify its representation. That is, we want to identify and implement the relevant features (*behaviors*) of components that mimic—as closely as possible—their real-world counterparts. (Please recall the banking example from Chapter 10.)

Encapsulation extends this idea:

Encapsulation is a design methodology wherein all implementation details (data and code) of a component remain *private*. Programmers only know—and interact with—the component's *public* interface (i.e., its *methods* and *functions*).

In other words, we want each component to remain a stand-alone module that provides a fixed, defined set of interfaces without exposing any of its implementation details.

This approach is not new, nor is it unique to computer science. We use tools and devices every day despite not knowing how they work.

---

[35] You may have heard or even used some of these products: Oracle's Database, IBM's DB2, Microsoft's SQL Server, etc.

Consider that most of us drive automobiles without understanding the inner workings of combustion engines, anti-lock brakes, or rack-and-pinion steering. Moreover, every car manufacturer designs and implements these components uniquely. Yet, through their public interfaces—brake pedals, accelerator pedals, and steering wheels—we drive cars without a second thought. Also, because they are independent, we can replace/upgrade components (e.g., brakes) without affecting any other subsystem.

Similarly, if we carefully employ *abstraction* and *encapsulation* in software design, other programmers may use the components we develop with minimal instruction. Moreover, we may replace/upgrade modules without affecting any other system component. (We'll see more examples of *abstraction* and *encapsulation* in Chapter 12.)

System Development Phase

Programmers begin building the application during the System Development phase. For computer geeks like myself, this is the best part of a project because we see all the moving parts start to come together: the architecture comes to life, code progresses from crawling to running, and disparate components begin working together.

However, developers do much more than write code in this phase. Industry estimates vary, but developers spend less than 50% of their time writing code. Some of their other responsibilities include:

- Complete low-level designs of their components

- Reconcile any design or component interface inconsistencies

- Review their code with peers to ensure quality

- Develop tests to demonstrate the accuracy of the code they write

- Fix any bugs that testing uncovers

- Prepare documentation to assist future programmers who might have to change the code (i.e., to fix bugs or add/modify features)

A Day in the Life of a Programmer

A typical workday for a programmer might look something like this:

- Review emails from other developers: Communication is critical in large development projects. Other programmers might need questions answered before they can continue coding some aspect of the system.

- Correct all errors found in overnight testing: In many IT shops, automated tests run nightly to determine if any "new" code (i.e., code written the previous day) has "broken" any existing code.[36] (This happens regularly during the early coding phases of projects.) Developers must fix these issues quickly because they can affect the testing of other programmers' code.

---

[36] We call this Automated Regression Testing.

- Write technical documentation: Depending on the methodology adopted and organizational needs, some systems require volumes of documentation. In some shops, developers must memorialize their designs, coding strategies, and testing techniques.

- Consult with users, business analysts, and supervisors: Despite the precision inherent in the underlying hardware, software development is an innately human undertaking. Therefore, it's subject to all the frailties, imperfections, and idiosyncrasies one might expect with such endeavors. Consequently, despite everyone's best efforts, questions, ambiguities, and inconsistencies arise daily in most large projects: requirements are often vague or ambiguous; designs are sometimes incomplete or imprecise; users frequently vacillate on the features they'd like included in their system. Programmers must raise these issues promptly, and responsible parties must provide timely solutions. Otherwise, large segments of the project may grind to a halt awaiting their resolution.

- Write code: This is a developer's *raison d'être*. It's at once creative, challenging, frustrating, disappointing, maddening, and rewarding. Most programmers love their jobs, and although they frequently change companies,[37] most remain in the profession in some capacity throughout their careers.

Despite an advertised forty-hour workweek, most programmers pound their keyboards for more than eight hours a day. The reasons are many: estimating is a black art; code doesn't behave as expected; and, well, life happens.

It often takes long hours and individual perseverance to overcome the seemingly endless list of requisite, unexpected obstacles. Moreover, as a project approaches its due date, the days grow even longer, and the pressure increases exponentially.

System Integration Phase

Once a "critical mass" of components becomes available, the development team begins combining them into a cohesive application. We refer to this process as *System Integration*. This procedure ensures that all in-house developed modules, along with any purchased components, function as a unified whole.

System Integration is often a complicated task and requires a dedicated individual, called a *System Integrator*, to manage it. As part of this effort, the System Integrator typically develops a packaging methodology allowing for simple delivery and deployment of the software into target execution environments. You've seen this type of packaging at work whenever you download and install applications on your phones and home computers.

Functional Testing Phase

As a project nears completion, the System Integrator prepares the application for *Functional Testing*. During this phase, an independent set of engineers attempt to "break" the code. That is, they run a suite of tests deliberately engineered to cause failures. Though testers

---

[37] Programmers frequently change jobs to get an opportunity to work with the newest technologies.

don't always garner the "love" that developers receive, their mission requires as much ingenuity and creativity as software development. Their efforts ensure that the application accurately delivers all the functionality specified in the requirements.

The testing team informs development staff whenever they uncover potential anomalies. The programmers must then "debug" the problem. Specifically, they must identify the underlying source of the issue[38] (called the *root cause*) and fix it (see the next section).

All aspects of system development are human endeavors; mistakes are therefore inevitable and unavoidable. Nonetheless, bug reporting often causes undue friction within IT shops. Developers challenge the testers and their results. Management shudders at the very thought that there are problems because of the time and costs involved in mitigating them. And users are often overheard bemoaning the incompetence of developers.

Due to this ongoing tumult, test engineers believe they are underappreciated and repeatedly feel compelled to remind everyone that they are only "doing their jobs." For myself, I always felt grateful when testing uncovered one of my bugs. Every error found in testing was one less that users would experience; this saved time, money, and—most important to me—personal embarrassment.

Debugging

Regardless of how or when software bugs rear their ugly heads, developers must address them. Although some issues are trivial to identify and fix, others are quite vexing. The latter type can be the bane of a programmer's existence; below are some examples:

- User-driven: A feature works as advertised for user A but fails with user B.

- Data-driven: The system functions appropriately when ordering product A but fails when ordering product B.

- Processing Sequence: The system only fails when executing a set of tasks in a specific order. For example, the application correctly processes task A, followed by task B. But it fails when the steps execute in reverse order. (This is one of the most difficult bugs to find.)

- Time of Day: The system operates correctly for the day shift but fails for night shift users.

The most common way developers identify a bug's root cause is by attempting to recreate the failure scenario in the lab under controlled conditions. By doing so, programmers can introduce instrumentation into the system that identifies and isolates the issue.

Unfortunately, this approach is not always successful. In rare cases, programmers require more invasive techniques to pinpoint an issue. For example, development teams might release a version of the application that includes additional instrumentation that generates "private" diagnostic messages that (hopefully) identify the underlying problem.

---

[38] We call this *walking the cat backward*.

We try to avoid this approach because it's expensive and may negatively affect system performance and user experience.

Once developers identify a bug's root cause, they must remedy it; this usually requires a coding change.[39] As a safeguard, whenever programmers modify code, they must retest the application to ensure that the fix works and doesn't undermine any other part of the system. That is, we must verify that the repair doesn't introduce a new bug.[40]

Performance Testing Phase

As its name might imply, the Performance Testing phase ensures that the system meets (or exceeds) its performance requirements. Performance testing identifies such issues as system bottlenecks (processing chokepoints), poorly performing components, network congestion, etc.

To accomplish this, testing staff employs tools that simulate many users accessing the system simultaneously. During testing, the team measures hardware utilization, network bandwidth consumption, and application responsiveness. Like functional anomalies (i.e., bugs), designers and developers must understand the root cause of performance issues and engineer resolutions when they arise.

Unfortunately, however, there are cases when poor performance is the result of significant design deficiencies. In such situations, the repercussions and recriminations ripple-like shockwaves through the project team. System redesigns are costly because of the amount of code that might require rewriting.

Professional development teams conduct performance testing as early as possible during the project lifecycle to prevent such scenarios. The sooner you know, the easier problems are to fix.

Acceptance Testing and Deployment

After completing functional and performance testing, systems typically undergo a round of *Acceptance Testing*, allowing users a "test drive" of the application. This step is often a formality: most professional project managers will not let users near a system until they're confident it's "ready for prime time."

Once Acceptance Testing concludes,[41] the project manager declares the system "ready for production." For custom-built, in-house applications, System Integrators will then make the application available to the intended user community (e.g., they replace the "old" payroll system with the "new" one). For apps such as those you might load onto your phone or PC, it means that the product is ready for downloading.

After product deployment, the project team engages in the most significant phase of the software development lifecycle: the Release Party. This is when the entire development staff gather to eat pizza, drink beer, share nerdy jokes, and complain about, well, everything related to the project.

---

[39] This is not always the case. Sometimes the requirements are wrong, or erroneous test scenarios generate "false positives." In addition, hardware failures or environmental issues can cause an apparent "application" bug. In such instances, coding changes are not necessary to mitigate the problem.

[40] This scenario happens quite often; we call it *break on fix*.

[41] Hopefully without uncovering any additional bugs—it's very embarrassing when that happens!

Maintenance Cycles

Like complex machinery, large-scale software systems require periodic maintenance to fix minor bugs, add new features, tweak performance, etc. IT shops usually treat a maintenance release as a mini development cycle, and it will typically progress through all phases of the SDLC.

ADVANCED TOPIC: DEVELOPMENT METHODOLOGIES

In the previous sections, we reviewed the most important tasks required to standup a typical software application. However, some complex systems like self-driving vehicles, SpaceX rocket launches, and Air Traffic Control systems may take years (even decades!) to complete. Moreover, lives depend on them.

This begs the obvious question: given such high stakes, how can we manage projects of this magnitude and consequence in a manner that ensures their success despite the inevitable design flaws, bugs, requirement changes, personnel turnover, etc. In short, even though "life happens," how do we ensure nothing terrible happens?

Throughout its history,[42] the IT industry has established various methodologies to manage the development of complex systems. While they may vary in approach and degree of formality, each offers a framework under which development teams can identify, understand, and manage a project's goals, tasks, schedules, and interdependencies.

The sections that follow present some of the more commonly used SDLC methodologies. However, please keep in mind that they're not rigid and inflexible. When adopting one of them, project teams will often customize or refine it to meet their specific needs.[43]

Waterfall Model

One of the oldest methodologies, the Waterfall Model partitions development tasks into a linear set of dependent phases. As depicted in Figure 11.7, one phase does not begin until the prior one completes.

Each phase has a defined set of tasks and deliverables (e.g., documents, code, test scenarios, etc.) that serve as *Exit Criteria*: the project team must complete them before progressing to the next phase. Moving forward, the deliverables from one phase serve as *Entrance Criteria* (i.e., input) to the next.

The Waterfall Model is intuitive, simple to understand, and identifies every task required to complete each phase. However, due to its linear nature, this methodology suffers several shortcomings. One of the most glaring is that it defers testing until the later part of the development cycle. Thus, project teams might not uncover critical deficiencies—such as design flaws or poor performance—until it's too late to mitigate them without impacting schedules and budgets. This model is also inflexible when adapting to the all-too-frequent course alterations that occur with large projects.

---

[42] Please keep in mind that the first commercial computers only became available in the early 1960s. Thus, compared to other disciplines, Computer Science is still in its infancy and, as a result, the IT industry is still in search of the perfect methodology.

[43] In my experience, I've found such adaptations to be both good and bad—mostly bad.

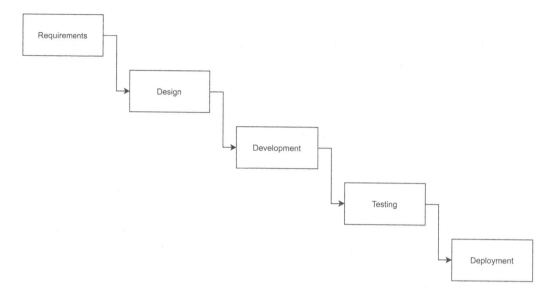

FIGURE 11.7 The Waterfall Model.

Prototyping

Prototyping is a development methodology most often used when the project team is not clear on requirements or design. In such cases, developers construct small segments of the system for users to review and evaluate. Then, based on the feedback they receive, programmers can revise and extend the prototype as needed. The project team repeats this process until garnering sufficient data to design and build the final product with confidence (see Figure 11.8).

Prototyping encourages alterations to requirements and designs as the team acquires incremental knowledge about the solution; this methodology invariably leads to a quality product. (There's no substitute for hands-on experience.)

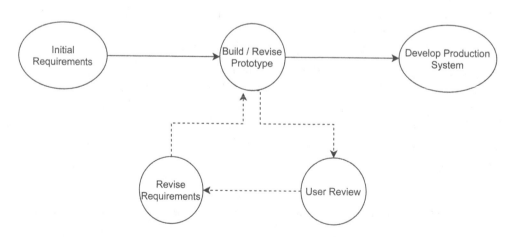

FIGURE 11.8 Prototyping methodology.

Nonetheless, this approach can become quite costly. Much of the early code is not production-caliber and is often unusable in the final product. It, therefore, needs rewriting—almost doubling development costs. Moreover, unless tightly monitored, schedules can extend indefinitely due to the team's desire to refine and perfect the prototype.

Agile Methodology

As the software industry came of age, linear methodologies (such as the Waterfall Model) dominated project management. Due to the complex nature of application development, the IT community believed that formal and rigid development practices led to timely, quality products. And they did—except when they didn't.

As noted above, one of the shortcomings of linear methodologies is that project teams often discover issues late in the game. For example, because design activities complete long before coding begins, users might not uncover shortcomings in screen layouts until the testing phase.

At this point, budgets might rule, and subpar functionality might find its way into production because the business might not have the time, money, or appetite to undertake the wholesale changes required to correct the problems. Burdened with a "custom" system they believe is less than adequate, users grow frustrated with the software team. (*Why can't programmers ever get it right?*) In turn, the development staff feels unappreciated because they built the system exactly "to spec." (*The users are never satisfied!*) Although prototyping methodologies often avoid such results, many organizations can't underwrite the costs and time needed for a "trial and error" approach.

Unfortunately, such outcomes were all too common—until the 1990s, when "lightweight" development methodologies began to emerge and evolve. One of the most prominent, called *Agile*, is a hybrid of the linear and prototyping approaches.

Under the Agile method, users and developers work collaboratively to divide projects into small increments of work called *sprints*, during which the team completes all tasks: requirements specification,[44] design, coding, and testing (see Figure 11.9). Each sprint aims

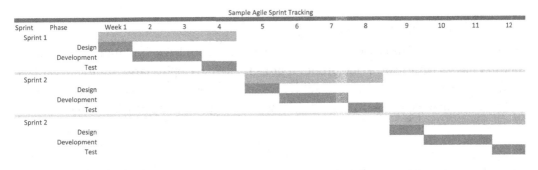

FIGURE 11.9   Simplified example of Agile Sprint Tracking.

---

[44] In Agile, requirements take the form of *User Stories*.

to deliver a small "chunk" of fully functional and well-tested software into production. In this manner, applications grow incrementally rather than deploying as a "big bang."

During each sprint's design phase, users collaborate closely with developers to ensure the product will meet the organization's needs. Such immediate feedback allows the development team to correct issues quickly.[45] Moreover, if requirements change due to shifts in market or business conditions, Agile teams are, well, agile, and can respond rapidly.

Some of the advantages of the Agile methodology include:

- It's flexible and adaptive

- Fosters a team approach

- Improves product quality[46]

Some of Agile's disadvantages include:

- Sprint teams create little or no documentation—this can adversely affect future development.

- Sprints can generate feature *backlog* (i.e., accrued functionality that can't fit into a *sprint*) and *technical debt* (accrued coding modifications triggered by design changes). If not addressed promptly, *backlog* and *technical debt* may undermine application design and slip project schedules.

- It's often the case that when an Agile project "kicks-off," the technical staff receives little time to develop an initial architectural design. Moreover, like its prototyping counterpart, the architecture evolves with each sprint as developers understand more about the application. Nonetheless, because they want to deliver as much business functionality as quickly as possible, users rarely apportion enough time in *sprints* for developers to restructure existing code to comply with changing designs.[47] This approach often leads to "cracks in the foundation."

Software Development Challenges

Despite their intent, all IT project management techniques have limitations. To date, the industry has yet to develop a "foolproof" approach[48] for developing large-scale applications.

Moreover, despite conscientious adherence to a methodology, most software projects still fail.[49] Nonetheless, regardless of their shortcomings, most applications do eventually

---

[45] We refer to this as "fail fast." Project teams like to "fail fast" because it simplifies and reduces the cost of coding changes.

[46] Unfortunately, there's little empirical evidence in support of this claim.

[47] We refer to this as *refactoring*.

[48] When striving for the perfect solution, one should remain mindful of one of Murphy's laws: *Nothing is foolproof to a sufficiently talented fool.*

[49] Some studies estimate that between 70% and 80% of all software development projects fail.

find their way into production. Thus, failure in a software development context constitutes any of the following:

- Feature: The application never meets user expectations

- Financial: Development costs exceeded projected budgets

- Technical: Delivered software is not of sufficient quality

- Catastrophic: The development team couldn't deliver some (or all) of the required features because of unrealistic expectations or technological limitations

The reasons why software projects fail are, unfortunately, too numerous to discuss in this book. However, based on my experience, these are the most common:

Poor Requirements In all my years in practice as a software architect, this is by far the most common cause of project failures. When requirements are missing, unclear, or ambiguous, developers—depending on their maturity and professionalism—will respond in one of the following ways:

- Ignore them—they won't write any code at all

- Create their own interpretations—which might be incorrect

- Ask for advice and clarification—always the best choice

One of the main reasons why requirements are inadequate is because many users believe they can write them. Unfortunately, most can't—and the project suffers accordingly.[50]

Knowledge Siloes In most organizations, every department jealously guards its information. After all, knowledge is power, and it's human nature to protect one's turf. Unfortunately, this does not play well when developing large-scale systems. Like accountants and lawyers, developers need to understand the business they're automating in as much detail as possible. Groups that don't share and don't "play nicely in the sandbox" significantly affect a project's success—whether they are aware of it or not.[51]

---

[50] Please keep in mind that it's typically the users who "write the checks." Thus, when applications fail because of poorly written requirements, project managers invariably direct the finger of blame toward the development team, not the people with the money. This is the bane of a developer's existence; most of us just deal with it and press on.

[51] Yes, it's the development team who'll bear the blame in these situations as well.

Unrealistic Schedules   I shouldn't admit this, but almost every project I've worked on has been late. This can be due to human error, technological challenges, unexpected issues, or changes in business direction. However, much of the time, it's caused artificially by zealous project managers who seem compelled to establish aggressive schedules during a project's initial planning phase.

As previously mentioned, software development is more of an art than a science. Moreover, every line of code ever written is for the first time. Nonetheless, it's difficult for some folks to comprehend that it's tantamount to impossible to estimate how long something will take to complete if it's never been done before!

That notwithstanding, seasoned software professionals, guided by experience, can offer reasonable and predictable estimates for development projects. Unfortunately, project managers rarely ask the professionals for input during a project's formative phases (when they establish the budgets and schedules). Moreover, in the rare case when project managers do ask for assistance, they seldom afford software professionals enough time to conduct their due diligence.[52] Thus, I've developed the following scheduling rule of thumb: start with your initial estimate, double it, and then add 20%.

Personnel Turnover   Software development requires a significant amount of knowledge across several disciplines: technology, business, and organization. It requires time for developers to acquire the necessary experience and expertise. When projects lose key contributors—due to retirement, resignation, promotion, transfer, etc.—quality and schedule will suffer.[53] In my experience, personnel turnover is also a result of what I call "technical candy." Most developers want to use the newest technologies and are willing to change jobs frequently to get their hands on them.

Integration Challenges   It never ceases to amaze me how even seasoned developers expect disparate software components to come together seamlessly when integrating them for the first time. Things rarely unfold as planned in the world of IT. There are just too many variables. Software integration takes time and effort, so anticipate some "speed bumps" and plenty of frustration.

---

[52] Yep, you're right. Even in this case it's the development team's fault when projects don't complete on schedule.

[53] Of course, there's also *addition by subtraction*: I've worked on many projects that were better off when managers "encouraged" poorly performing individuals to move on.

Inadequate Testing Unfortunately, when it comes to quality, many project team members forget that system testing serves as a project's last line of defense. Consequently, test engineers rarely receive the respect and recognition they deserve.[54] For example, when schedules are in jeopardy, many project managers will shorten testing cycles rather than miss a delivery date. Also, when budgets are tight, testing groups often must "scrounge" for equipment and resources. This mindset is short-sighted: the more you test, the better the product. It's that simple.

Although they may be far less common, there are many other reasons why software projects fail. Please keep in mind that, in the end, you're dealing with both human beings and complex technology: a volatile and complicated combination.

SUMMARY

In this chapter, we've reviewed classes of programming languages, program execution, and the software development lifecycle. We've also discussed how complicated it is to develop large systems and presented some of the methodologies the IT industry uses to mitigate against failure.

In Chapter 12, we'll combine all the knowledge we have acquired thus far to write a complete, functioning computer program. I don't know about you, but I'm looking forward to it.

---

[54] In some cases, however, the lack of respect is warranted. Project teams frequently "hide" underperforming personnel in testing groups. In addition, many organizations that readily invest scads of money educating developers won't spend a dime training test engineers.

CHAPTER 12

Putting It All Together

A pessimist sees the difficulty in every opportunity; an optimist sees the opportunity in every difficulty.

WINSTON CHURCHILL

INTRODUCTION

We've covered a lot of material thus far—from digitization to compilation. Now it's time to put this new-found knowledge into practice and develop a working program from beginning to end. However, before getting started, there are some decisions we must make.

First, what programming language should we use? There are many viable choices: C, C++, Python, Java, C#, and Visual Basic, to name a few. Any one of them would serve our purpose.

After some reflection, I opted to use Java for the following reasons:

• The language is readily available, and it's free to use

• If we avoid its advanced features, Java is, relatively speaking, easy to understand

Second, what type of program should we build? The challenge is selecting an example that's small enough to get our heads around yet big enough to demonstrate how professional developers ply their trade.

After some additional deliberation, I decided upon a project I've often used when teaching programming to undergraduate students. We'll write a program that converts Roman Numerals into their integer equivalents. This project is challenging enough to make it interesting, yet it requires only a simple user interface—developing a full graphical user interface (GUI) would be tedious. (As an aside, don't be concerned if you've forgotten how to work with Roman Numerals; we'll have a quick refresher course shortly.)

As a bonus, after we complete this program, we'll write another that does the opposite: converts integers into Roman Numerals. Finally, in the Advanced Section, we'll discuss ways to combine these two programs into one. Trust me; this will be cool.

General Comments

Before we begin, I'd like to make some general comments regarding the overall approach we'll adopt.

Coding Style When writing code, software professionals have two audiences: the compiler (or interpreter) and other programmers. Developers should write code that functions correctly and is easy for other programmers to comprehend. (We refer to this latter attribute as *readability*.)

Unfortunately, readability is somewhat subjective, and there are raging debates in the industry regarding *coding style*—the presentation guidelines developers should employ when writing code in a specific language. The disputes involve such items as the positioning of syntactical elements such as braces and parentheses, how frequently to include comments, and where one should insert *whitespace* (i.e., blank lines, tabs, and spaces to enhance readability).

For pedagogical clarity, I opted for a presentation style that allows readers to focus on the code's essential aspects.

For those of you new to software development, you can safely ignore this discussion. For readers with some programming experience, I ask you to look past any objections you might have with the coding style I adopted for this book.

Coding Clarity The examples presented below serve a pedagogical purpose and are therefore not necessarily representative of professional programming practices. More to the point, this text is not a Java tutorial, nor is it a discourse on writing efficient code. Thus, in most cases, I opted for clarity rather than cleverness. That said, I make no apologies for the approach I've adopted—I don't want readers of this book growing frustrated trying to weave their way through complex language-specific constructs.[1]

Comprehension Despite their pedagogical simplicity, the examples still embody a significant level of complexity. Readers new to coding shouldn't feel concerned if they don't understand every language nuance immediately; this is normal and expected. I assure you, if you focus on understanding how the programs work at a high level, the details will follow.

---

[1] In addition, I'm trying to minimize the number of unpleasant emails I might receive when some developers read the code.

Coding Environment   I developed the Java code for these examples using the Cygwin programming environment running atop Windows 10. In addition, I tested them running directly under Windows as well. Thus, if you type (or copy) the code accurately, the programs will run correctly and generate output identical to that in the examples.

However, there is one important caveat: When you type in the code, DO NOT include the line numbers that appear in the listings. I added them for discussion purposes only. The Java compiler will not treat you kindly if your program file contains them.

Development Methodology   To simplify the presentation, we've adopted a stream-lined development methodology that includes only the Requirements, Design, and Coding phases. We'll omit tasks such as preparing documentation, requirements analysis, developing test scripts, etc.

Requirement Specification   Today, it's common practice to express requirements as *User Stories*, which specify system functionality using everyday language. This approach, however, would be "overkill" for our purposes. Thus, we'll enumerate requirements using a simple numbered list format.

Now that we've dispensed with the disclaimers, let's get started.

Roman Numerals: A Quick Overview

Roman Numerals use letters to represent numeric quantities. Table 12.1 presents the basic set of characters and their corresponding values.

Reading left-to-right, we construct Roman Numerals by combining the letters—from highest to lowest—in varying combinations. For example:

$$M = 1,000; MC = 1,100; DC = 600; CV = 105$$

The same letter may appear in sequence at most three times. In such cases, we sum the value of the letters, as in:

$$X = 10, XX = 20; XXX = 30; LXX = 70; CCII = 202$$

However, to generate values that would require four letters (e.g., 40), we instead use a technique involving subtraction. Specifically, when we combine two different Roman

TABLE 12.1   Roman Numeral Values

Roman Letter	M	D	C	L	X	V	I
Integer Value	1,000	500	100	50	10	5	1

letters in reverse value order, they form what's called a *two-character sequence*. In all such cases, we subtract their values rather than sum them:

$$XL = 40\ (50-10 = 40);\ IV = 4\ (5-1 = 4);\ XC = 90\ (100-10 = 90)$$

Just to be clear, let's try one more example:

$$CCXCIV = 294$$

The leftmost character, "C", is followed by a letter of lesser or equal value (another "C"). This implies addition, and we compute an intermediate result of 200 (100 + 100 = 200).

However, the next two letters (XC) appear in reverse value order (low to high). Thus, they form a *two-character sequence* that indicates subtraction; the result is 90 (100 – 10 = 90).

The same holds for the remaining two letters (IV): they also form a *two-character sequence* that, in this case, results in the value 4 (5 – 1 = 4).

When we sum all the intermediate results, we arrive at a final numeric value of 294.

Given the above description, we can codify the rules of Roman Numeral construction as follows:

1. If the numeric value appears in the table, we use its corresponding letter. For example, if we want to express the value 50, we write "L." Similarly, we'd write "C" to represent the value 100.

2. If two Roman letters of the same or lesser value appear in sequence, we sum the corresponding values. For example:

   - VI = 6 (5 + 1 = 6)

   - CCII = 202 (100 + 100 + 1 + 1)

   - MMXX = 2020 (1000 + 1000 + 10 + 10)

3. Whenever a letter of lesser value appears before one with a greater value, they form a *two-character sequence*, and we subtract their values. For example:

   - IV = 4 (5 – 1 = 4)

   - XL = 40 (50 – 10 = 40)

   - CD = 400 (500 – 100 = 400)

4. We can combine the above rules to express arbitrary values, as in:

   - MCMXCIX = 1,999 (1,000 + (1,000 – 100) + (100 – 10) + (10 – 1) = 1,999)

Although there are variants and extensions to these basic rules, we'll use only the standard forms described above for our program.

PART 1—CONVERTING ROMAN NUMERALS TO INTEGER

As mentioned previously, we'll adopt a streamlined methodology to develop our example programs. Specifically, we'll start by defining the requirements, then perform some analysis, design the program, and finally code and test it. Although the process is simplified, you should nonetheless gain significant insights into how IT professionals design, build, and test the software applications you use every day.

Let's get started.

Requirements

The requirements for our Roman-to-integer conversion program are as follows:

1. Upon invocation, the program prompts the user to enter a Roman Numeral.

2. After reading the user-entered value, the program performs the conversion and displays the result to the user.

3. The user may enter the Roman Numerals using either uppercase or lowercase letters.

4. The Roman Numerals entered by the user must adhere to the rules specified above.

5. The program will provide a simple line-oriented user interface. (As mentioned previously, developing a GUI would be too complicated and time-consuming.)

Design

After reviewing the requirements, we can summarize the tasks our program must perform:

- Initialize the execution environment. This entails setting up the variables (i.e., "storage lockers") that our program requires to process user input and compute the integer value. (We'll discuss the details when we review the code.)

- Prompt the user for input. To do this, the program will use a built-in Java utility that displays characters on the screen.

- Read the Roman Numeral that the user entered. As in the previous bullet point, the program will leverage a built-in Java function that reads a string of characters from the keyboard.

- Compute the integer value.

- Display the result to the user, again using a built-in utility.

Code

Based on the requirements and design discussed above, I wrote a program that performs the Roman Numeral to integer conversion. The Java code appears in Listing 12.1.

However, before we describe how the code works, I need to make a few general comments.

- I typed the code into a file named ROMANTOINTEGER.JAVA. You'll have to do this if you want to compile and execute this program on your system. (I'll demonstrate how later in the section.) However, as mentioned previously, the line numbers appearing in the listings are NOT part of the source code proper; I added them for notational convenience. Do not include line numbers if you create your own program file. If you do, the compiler will issue a series of error messages and abort the compilation.

- Double forward slashes ("//") are a notational convention Java supports that allow developers to include *comments* in their programs. They serve as notes/explanations for other programmers; the compiler ignores all characters beginning with "//" through the end of the line.

- Like comments, the compiler ignores blank lines. I insert them liberally in the listing to enhance the readability of the program.[2] Judicious use of *whitespace* (i.e., spaces, tabs, and blank lines) improves a program's *readability*, helping other programmers comprehend the code.

- One final caveat: This text is not a Java tutorial. As such, there are a few language constructs that we will not discuss in detail. Not to worry; I assure you this will not affect your understanding of the code.

Program Listing
Listing 12.1 presents the program's code in its entirety.

LISTING 12.1 ROMAN TO INTEGER

```
001 //
002 // THIS PROGRAM CONVERTS ROMAN NUMERALS TO INTEGERS
003 //
004
005 IMPORT JAVA.UTIL.SCANNER;                    // REQUIRED TO USE JAVA'S SCANNER LIBRARY
006
007 PUBLIC CLASS RomanToInteger              // NAME OF THE CLASS/PROGRAM
008 {
009     //
010     //     THIS FUNCTION WILL CONVERT A ROMAN
011     //     CHARACTER INTO ITS EQUIVALENT INTEGER VALUE
012     //
013     PUBLIC STATIC INT convertRomanCharToIntegerValue ( CHAR romanChar)
014     {
015             //
016             //     DETERMINE THE CHARACTER AND
017             //     RETURN THE APPROPRIATE VALUE
018             //
019             SWITCH ( romanChar )
020             {
021                     CASE 'I':
022                             RETURN ( 1 );
023
024                     CASE 'V':
```

---

[2] There are many attributes that, collectively, constitute "professional-caliber" code. Though subjective in nature, *readability* is one of them.

```
025                               RETURN( 5 );
026
027                  CASE 'X':
028                               RETURN( 10 );
029
030                  CASE 'L':
031                               RETURN( 50 );
032
033                  CASE 'C':
034                               RETURN( 100 );
035
036                  CASE 'D':
037                               RETURN( 500 );
038
039                  CASE 'M':
040                               RETURN( 1000 );
041
042                  DEFAULT:
043                               //
044                               // BAD CHARACTER - FLAG AS ERROR
045                               // INFORM THE USER
046                               //
047                               SYSTEM.OUT.PRINTLN( "INVALID ROMAN CHARACTER: '" +
                                 ROMANCHAR + "'" );
048                               RETURN( 0 );
049            }
050     }
051
052     //
053     //      THIS FUNCTION CONVERTS ROMAN NUMERALS TO INTEGERS
054     //      THIS IS WHERE THE 'HEAVY LIFTING' TAKES PLACE
055     //
056     PUBLIC STATIC INT CONVERTROMANSTRINGTOINT( STRING INPUTSTRING )
057     {
058            //
059            //      DECLARE VARIABLES ("STORAGE LOCKERS")
060            //
061            INT I = 0;
062            INT INTVALUE = 0;
063            INT ROMANVALUE = 0;
064            INT NEXTROMANVALUE = 0;
065
066            //
067            // STEP THROUGH EVERY CHARACTER IN THE STRING
068            //
069            FOR( I = 0; I < INPUTSTRING.LENGTH(); I++ )
070            {
071                   //
072                   // GET THE VALUE OF THE CURRENT CHARACTER
073                   //
074                   ROMANVALUE = CONVERTROMANCHARTOINTEGERVALUE( INPUTSTRING.
                       CHARAT(I) );
```

```
075
076                    IF( I < INPUTSTRING.LENGTH()-1 )  // ARE THERE MORE CHARACTERS?
077                    {
078                           //
079                           //    THERE ARE MORE CHARACTERS -
080                           //    GET THE INTEGER VALUE OF THE 'NEXT' CHARACTER
081                           //
082                           NEXTROMANVALUE = CONVERTROMANCHARTOINTEGERVALUE(
                              INPUTSTRING.CHARAT(I+1) );
083
084                           IF( ROMANVALUE >= NEXTROMANVALUE )
085                           {
086                                  //
087                                  //    THIS IS THE 'NORMAL' CASE
088                                  //    E.G., 'VI' OR 'LX'
089                                  //    WE CAN IGNORE THE 'NEXT' VALUE AND
090                                  //    ADD THE CURRENT VALUE TO THE RUNNING
                                             TOTAL
091                                  //
092                                  INTVALUE += ROMANVALUE;
093                           } ELSE IF( ROMANVALUE < NEXTROMANVALUE ) {
094                                  //
095                                  //    THIS IS THE CASE WHEN WE HAVE TO
                                             SUBTRACT VALUES
096                                  //    E.G., 'IV' OR 'XL'
097                                  //    IN THIS CASE WE HAVE TO SUBTRACT THE
                                             'CURRENT' VALUE
098                                  //    FROM THE 'NEXT' VALUE AND ADD THAT RESULT
099                                  //    TO THE RUNNING TOTAL.
100                                  //
101                                  INTVALUE += (NEXTROMANVALUE - ROMANVALUE);
102
103                                  //
104                                  //    WE ALSO HAVE TO 'CONSUME' THE 'NEXT' VALUE
105                                  //    BECAUSE WE'VE ALREADY PROCESSED IT.
106                                  //
107                                  I++;
108                           }
109                    } ELSE {
110                           //
111                           //    NO MORE CHARACTERS AFTER THIS ONE ...
112                           //    ADD THE LAST CHARACTER TO RUNNING TOTAL
113                           //
114                           INTVALUE += ROMANVALUE;
115                    }
116             }
117
118      //
119      //    RETURN THE RESULT
120      //
```

```
121                 RETURN ( INTVALUE ) ;
122     }
123
124     //
125     //      EXECUTION BEGINS IN THE 'MAIN()' FUNCTION
126     //
127     PUBLIC STATIC VOID MAIN ( STRING [] ARGS )
128     {
129             //
130             //      DECLARE VARIABLES ("STORAGE LOCKERS")
131             //
132             INT INTVAL;
133             SCANNER   USERINPUT;
134             STRING    ORIGSTRING;
135             STRING    ROMANSTRING;
136
137             //
138             //      THE LINE OF CODE PREPARES THE SYSTEM
139             //      TO ACCEPT INPUT
140             //
141             USERINPUT = NEW SCANNER ( SYSTEM.IN ) ;
142
143             //
144             //      PROMPT THE USER TO TYPE IN THE STRING
145             //
146             SYSTEM.OUT.PRINT ( "PLEASE ENTER THE ROMAN NUMERAL YOU WANT CONVERTED: " ) ;
147
148             //
149             //      READ WHAT THE USER TYPED
150             //      STORE IT IN 'ORIGSTRING'
151             //
152             ORIGSTRING = USERINPUT.NEXTLINE () ;
153
154             //
155             //      CONVERT INPUT TO LOWERCASE
156             //
157             ROMANSTRING = ORIGSTRING.TOLOWERCASE () ;
158
159             //
160             //      CALL THE FUNCTION THAT WILL PERFORMS THE CONVERSION
161             //
162             INTVAL = CONVERTROMANSTRINGTOINT ( ROMANSTRING ) ;
163
164             //
165             //      DISPLAY THE RESULTS TO THE USER
166             //
167             SYSTEM.OUT.PRINTLN ( "THE INTEGER EQUIVALENT OF: '" + ORIGSTRING + "' IS:
                " + INTVAL ) ;
168     }
169 }
```

The first three lines of Listing 12.1 are comments; the fourth is a blank line. (As stated previously, the compiler ignores these elements.)

The fifth line, IMPORT JAVA.UTIL.SCANNER;, is a *declaration* that instructs the compiler to load a *package* called JAVA.UTIL. Java *packages* contain prewritten code that we can include in our programs and use as if we wrote it.[3] In this case, we're using a utility called, SCANNER, which simplifies the tasks required to read characters entered by the user. (Below, we'll see how the program uses this package to read the Roman Numerals typed by the user.)

Lines 7 and 8 (PUBLIC CLASS ROMANTOINTEGER {) declare a class called ROMANTOINTEGER, which serves as our program name. This name is not "magic;" I chose it. As a result, however, Java requires that the source code resides in a file named ROMANTOINTEGER.JAVA.

Java Methods

Before we continue with the listing, let's take a small detour and review a Java construct called a *method*. (This section expands on the material previously presented in Chapter 10.)

Let's begin with a definition:

> A *method* is a named group of statements that collectively perform a specific function.[4]

In other words, we combine multiple statements into a single entity and execute them by name. This concept should be familiar to most of you. Consider the following examples.

- Most folks begin their mornings with a "daily routine" (e.g., wash your face, brush your teeth, make coffee, etc.). Although not "electronic," this is an example of a named set of tasks or instructions.

- When you launch a program (like a browser), you do so by name (or icon), and all the instructions execute.

- Anyone who has configured a routine on a smart device (e.g., Amazon's Alexa or Google's Smart Speaker) has created a method. For example, many users of these devices set up a "goodnight" routine that locks doors, adjusts thermostats, and arms their security system. Then, when they say "goodnight" to their smart device, all these tasks "execute."

When coding, programmers create methods that group instructions into logical units. Let's say we need to write a program that computes the average of two numbers. We could write the following Java code:

```
AVERAGE = (VALUE1 + VALUE2) / 2;
```

---

[3] This is an example of *code reuse*.

[4] Collections of instructions have many names besides *method: function, subroutine,* and *procedure.*

Alternatively, we could create a reusable method that computes the average of any two numbers as follows:

```
PUBLIC INT COMPUTEAVERAGE( INT v1, INT v2 )
{
        INT ANSWER;

        ANSWER = (v1 + v2) / 2;
        RETURN( ANSWER );
}
```

Whenever we wanted to use the method, we would invoke it as follows:

```
// OTHER CODE HERE ...
AVERAGE = COMPUTEAVERAGE( VALUE1, VALUE2 );
// MORE CODE HERE ...
```

Although trivial, this example demonstrates how methods encapsulate code and permits programmers to execute them as often as necessary:[5]

However, the suite of methods available to Java programmers extends well beyond the ones they write themselves. The Java language includes many *packages* (like SCANNER above), each of which contains numerous methods.

Moreover, there are countless *shareware* (free to use with some restrictions) and licensed, fee-based packages available for integration into your application. As a rule, before you begin programming a piece of code, it's often beneficial to conduct some due diligence to determine whether someone has already written it.

The MAIN() Method

Returning to the program, let's jump ahead to line 127 of the listing. We'll return to the code we skipped over shortly.

As you might expect, every program must have a starting point. By convention, Java programs begin execution with the first instruction contained in a method named MAIN().

In our program, the declaration for MAIN() appears at line 127:

```
PUBLIC STATIC VOID MAIN( STRING[] ARGS )
```

For our purposes, we can ignore the Java keywords, PUBLIC, STATIC, and VOID.[6] We can also safely ignore the code appearing between the parentheses, STRING[] ARGS.[7]

---

[5] Please note that professional programmers would not write such a function. I created this example for pedagogical purposes only.

[6] These keywords are part of what's called a *method declaration* in Java.

[7] We'll discuss *parameters* in the next section.

Thus, we can jump ahead and resume our discussion of the code beginning with lines 132 through 135.

```
INT          INTVAL; 8
SCANNER      IN;
STRING       ORIGSTRING;
STRING       ROMANSTRING;
```

As you may recall from Chapter 5, programs need variables (i.e., "storage lockers") to hold data. In Java, we use *declaration* statements to name and reserve storage for variables. Our MAIN() method requires four variables:

INTVAL   This variable contains the computed integer value.

USERINPUT   Of type SCANNER, the variable, USERINPUT, provides access to the keyboard

ORIGSTRING   The program uses this variable to store the characters entered by the user (i.e., the Roman Numeral the user wants converted)

ROMANSTRING   This is a copy of the user's input converted to lowercase. (The reason I use this variable will become clear shortly)

As we proceed, we'll understand how the program uses these variables to complete the conversion. Let's press on.

After some comments (lines 137–140), the first executable statement appears on line 141; this is where processing begins.

```
USERINPUT = NEW SCANNER( SYSTEM.IN );
```

This statement creates an instance of a built-in SCANNER *object* and assigns it the name USERINPUT. (For now, we can ignore the details and think of an *object* as a complex type of *variable*.[9]) We'll see how the program uses USERINPUT to read characters from the keyboard in short order.

On line 146, we use another built-in Java utility to prompt the user.

```
SYSTEM.OUT.PRINT("PLEASE ENTER THE ROMAN NUMERAL YOU WANT CONVERTED:");
```

When this statement executes, the user will see the text "Please enter the Roman Numeral you want converted:"[10] appear on the screen.

---

[8] This odd-looking spelling is an example of a naming convention called *Camel Case*. Used to enhance readability, Camel Case is the custom of combining multiple words into a single syntactical unit (i.e., a variable or method name). We indicate the beginning of each embedded word using a capital letter.

[9] For programming purists, an *object* is an *instance* of a *class* which is a complex *data structure* that contains both *data* (i.e., "storage lockers") and *behaviors* (i.e., *methods*).

[10] The quote (") characters will not appear on the screen. However, the blank character positioned after the colon (:) will "display."

After issuing the prompt, the program needs to read and save the Roman Numeral the user wants to convert. The code to do this appears on line 152.

```
ORIGSTRING = USERINPUT.NEXTLINE();
```

After this statement executes, the variable, ORIGSTRING, will contain every character entered by the user.

Before we move on to the next statement, I'd like to make two comments.

First, we've seen several instances of the keyword STRING, so it's time we define it. In Java,[11] a single character is of type CHAR. (The keyword, CHAR, is shorthand for *character*.) We refer to a sequence of CHARS (characters) such as "ABC", "$#&%", and "Hello World" as a STRING. Thus, by definition, a variable of type STRING is a collected sequence of CHARS.

So far, so good. Unfortunately, it gets a bit more complicated. A sequence of numerals, such as "123", is also a STRING. However, despite how counterintuitive it might seem, such strings don't have an intrinsic numeric value. As we'll see, to work with numeric strings in math operations, we must convert them to other data types such as INT (short for *integer*) or FLOAT (short for floating-point—i.e., any number that includes a decimal point as in "456.78").

Second, I used the variable, ORIGSTRING, to preserve the user-entered text. As you'll see, the program doesn't do much with this variable other than copying its value into another variable named ROMANSTRING. Strictly speaking, saving what the user entered is not necessary. However, I've learned that preserving data in its original form often helps when debugging errors—I do it as a matter of course.

Moving on, the next executable statement appears on line 157.

```
ROMANSTRING = ORIGSTRING.TOLOWERCASE();
```

The program uses a built-in STRING method, called TOLOWERCASE(), to convert all the characters the user entered to lowercase. There are a couple of reasons why I opted to do this.

As with natural languages, uppercase characters differ from their lowercase counterparts in Java.[12] Nonetheless, when working with Roman Numerals on paper, we might ignore case and print either "ccc" or "CCC" to represent the value "300."

Though obvious to humans, this can wreak havoc in a computer program. Why? Because "c" is a different character than "C" (please recall our keyboard discussion in Chapter 3).

Thus, if a program tests for an uppercase "C" and the user entered a lowercase "c", the program might not function as "expected" (i.e., we'd have a bug). The converse is true as well. If we test for a lowercase "c" and the user enters an uppercase "C", a similar problem will arise.

One way to avoid this issue is to have the program always test for both cases; but, this approach would needlessly complicate the code. Alternatively, the program could

---

[11] This is true for most programming languages.
[12] Please refer to the discussion of ASCII characters appearing in Chapter 3.

mechanically convert all user input to a single case; this simplifies coding because it eliminates the need to test for both.

As the code suggests, I elected to employ the conversion approach for this example. However, the choice to convert to lowercase was arbitrary; I could just as easily have opted to convert all user input to uppercase.

One last point. It may seem obvious, but developers should make life easy for the user, not themselves.[13] To that end, the approach we've taken allows users to enter Roman Numerals in either case.[14] Thus, in effect, that one statement—converting user-entered text to lowercase—offers two bangs for the buck: simplified coding and user convenience.

Only two statements remain in MAIN(). On line 162, the program calls a method that I wrote called CONVERTROMANSTRINGTOINT(). This function performs the actual conversion; we'll discuss its implementation in the next section.

The last statement, appearing on line 167, displays the result of the conversion to the user.

```
SYSTEM.OUT.PRINTLN ( "THE INTEGER EQUIVALENT OF: '" + ORIGSTRING + "' IS: " + INTVAL );
```

The built-in Java method, PRINTLN(), displays its arguments on the screen. For example, if the user entered the Roman Numeral "ccc", the program would generate the following output:

```
THE INTEGER EQUIVALENT OF: 'CCC' IS: 300
```

Note that the "+" operator allows us to "string together"[15] text and values to form a composite STRING.

One final comment. This type of high-level program design is quite common. We typically use MAIN() as a "driving routine" to coordinate processing but delegate all the "heavy lifting" to other methods. This approach fosters code reuse because we can package CONVERTROMANSTRINGTOINT() and CONVERTROMANCHARTOINTEGERVALUE() and make them available for use in other programs.

That completes our discussion of MAIN(). The next two sections describe how the program performs the conversion.

The CONVERTROMANCHARTOINTEGERVALUE() Method

Of the program's two remaining methods, CONVERTROMANCHARTOINTEGERVALUE() is smaller, so let's begin with it. Its declaration appears on line 13 of the listing.

```
PUBLIC STATIC INT CONVERTROMANCHARTOINTEGERVALUE ( CHAR ROMANCHAR )
```

---

[13] Sadly, many "professional" programmers don't seem to agree with my position on this.

[14] Astute readers might have observed that the program would also accept mixed-case input such as "CcC."

[15] The technical term for this operation is *concatenation*.

Again, we'll ignore the keywords PUBLIC and STATIC (we'll discuss the keyword INT below). But first, as promised in the previous section, we need to discuss the elements positioned between the parentheses: CHAR ROMANCHAR.

Java methods can accept input, execute code, and return a result.[16] Methods receive input via a *parameter list*. In our example, CONVERTROMANCHARTOINTEGERVALUE accepts only one (CHAR ROMANCHAR). However, you could write a function that accepts multiple parameters, as in the following example:[17]

```
INT ADDTHREENUMBERS ( INT N1, INT N2, INT N3 )[18]
{
        INT ANS;
        ANS = N1 + N2 + N3;
        RETURN ( ANS );
}
```

The function ADDTHREENUMBERS()accepts three integer values, computes their sum, and stores the result in a variable named ANS.

This example also demonstrates how Java methods can return values via a RETURN statement. As part of its declaration, the keyword, INT, indicates that ADDTHREENUMBERS()returns a value of *type* INT (integer) to its caller; the actual value is returned using a RETURN statement:

```
RETURN ( ANS );
```

We can invoke this method as follows:

```
SUM = ADDTHREENUMBERS ( 1, 2, 3 ); // RETURNS THE VALUE 6.
```

Returning to our program, we declare CONVERTROMANCHARTOINTEGERVALUE() to accept one argument—a single Roman letter (of type CHAR)—and return a value of type INT. When it executes, the method identifies the Roman letter it receives and returns its equivalent integer value.

Let's see how it works.

Using a Java SWITCH statement (beginning with line 19), the method compares the value it receives in ROMANCHAR with the various *case labels* appearing throughout its *switch block*.[19] When it finds a match (i.e., when ROMANCHAR matches one of the letters in the case labels), it executes the associated code.

---

[16] I should note for the sake of completeness that Java methods need not accept parameters nor are they required to return a value.

[17] In addition to varying in number, parameters may also differ in *type* (e.g., INT, CHAR, FLOAT, etc.).

[18] Once again, this example is for pedagogical purposes only; professional programmers would not code such a method.

[19] Note that the *switch block* contains only lowercase *labels* because of the case conversion we performed in the MAIN() method.

For example, if ROMANCHAR happen to contain the value "c" (thus matching the *case label* on line 33), it would execute the statement appearing on line 34:

RETURN (100);

The DEFAULT[20] label (line 42) is a catch-all: if none of the other labels match ROMANCHAR, the SWITCH statement executes the code beginning with line 47. (Note that we skipped the comment lines.)

For our program, if there is no match, the user entered an invalid Roman letter. In response, the method generates an error message and returns the value 0. (See line 48 in the listing.) Although not the best way to manage errors,[21] the example demonstrates that programmers must always assume the worst and code defensively.[22]

The CONVERTROMANSTRINGTOINT() Method

The heavy lifting in our program begins at line 56 with the declaration of the method CONVERTROMANSTRINGTOINT(). This function accepts one argument, INPUTSTRING, and its job is to convert that Roman Numeral into its equivalent integer value. As we proceed with this discussion, please keep in mind the conversion rules we reviewed earlier in the chapter.

Lines 61 through 64 contain variable declarations.

I	The method uses the variable, I, to track its progress as it steps through each character contained in the Roman string. (When reading through the code, please remember that we start counting at zero (0) in the IT world.)
INTVALUE	Serving as a running total, the variable, INTVALUE, contains the Roman string's integer value. Initially, CONVERTROMANSTRINGTOINT() initializes INTVALUE to zero (0). As it processes each Roman character, the method incrementally adds the appropriate integer value to INTVALUE's total. When processing terminates, INTVALUE contains the final value of the Roman Numeral the user entered.
ROMANVALUE	This variable contains the integer value of the Roman character the method is about to convert. (See conversion rule 2.)
NEXTROMANVALUE	This variable contains the *next* Roman letter's value and determines whether the method must process a *two-character sequence*. (See conversion rule 3.)

The code beginning on line 69 allows us to step through each Roman character in sequence. It's called a FOR loop, and we refer to the process as *iteration*.

---

[20] The DEFAULT label is optional in Java SWITCH statements.

[21] As a rule, it's preferable to have the "outer layers" of your program—the function MAIN() in our case—to validate data. It becomes increasingly difficult to respond to errors as you dive "deeper" into a program.

[22] In professional applications, developers spend a significant percentage of their time writing "protective" code.

A Java FOR loop is a control structure that programmers use to repeat a sequence of instructions a specified number of times.

As depicted in the skeleton code appearing in Listing 12.2, a Java *for loop* has two main parts: a *control section* and a *body*. Subdivided into three parts (separated by semicolons), the *control section* determines how many times the instructions appearing in the loop's *body* will execute. The loop's *body* may contain any number of valid Java statements.

LISTING 12.2 STRUCTURE OF A JAVA FOR LOOP

```
FOR ( INITIALIZATION; CONDITIONAL EXPRESSION; ITERATION EXPRESSION )  // CONTROL SECTION
{
        // BODY OF LOOP - ONE OR MORE VALID JAVA STATEMENTS
}
```

As a simple example, Listing 12.3 contains a FOR loop that displays the values 0 through 9. Processing begins with an initialization, INT I = 0;. This expression sets the loop's *control variable*, I;[23] to an initial value of 0; it executes only once just before the loop's first iteration.[24]

LISTING 12.3 AN EXAMPLE FOR LOOP

```
FOR ( INT I = 0; I < 10; I++ )
{
        SYSTEM.OUT.PRINTLN ( "THE VALUE OF I IS: " + I );
}
```

Before each iteration begins, the code evaluates the *conditional expression*, I < 10;. The *body* of the loop repeatedly executes while this expression yields a value of TRUE.[25] This test executes before *every* iteration, including the *first*: if the *conditional expression* doesn't evaluate to TRUE the first time, the loop's body won't execute at all.

The next component to execute is the FOR loop's *body*, which may include any number and type of valid Java statement (including nested FOR loops). Listing 12.3, the *body* contains only one statement which displays the value of I:

```
SYSTEM.OUT.PRINTLN ( "THE VALUE OF I IS: " + I );
```

---

[23] In general, this may be any appropriate value as it pertains to the iteration.
[24] The control variable may have any name; the name I is commonly used based on historical precedent.
[25] We call expressions that yield values of *true* or *false* Booleans or Boolean expressions (see Chapter 5).

After each iteration of the loop's *body*, the *iteration expression* executes. Typically, this statement "adjusts" the value of the *control variable* as appropriate. In this example, the *iteration expression* increments (i.e., adds 1 to the value of) the variable I:

```
I++
```

The output of Listing 12.3 appears in Listing 12.4.

LISTING 12.4   THE OUTPUT OF LISTING 12.3

```
THE VALUE OF I IS: 0
THE VALUE OF I IS: 1
THE VALUE OF I IS: 2
THE VALUE OF I IS: 3
THE VALUE OF I IS: 4
THE VALUE OF I IS: 5
THE VALUE OF I IS: 6
THE VALUE OF I IS: 7
THE VALUE OF I IS: 8
THE VALUE OF I IS: 9
```

The following template summarizes the processing of a Java FOR loop:

Execute the loop's *initialization expression* once (and only once)
Execute the *conditional expression* before every iteration—does it yield a value of TRUE? If so:
  Execute all the statements in the loop's *body*
  Execute the loop's *increment expression*
  Return to the *conditional expression*
When the *conditional expression* evaluates to FALSE, terminate the loop and resume execution with the statement immediately following the end of the FOR loop

---

Now, let's return to the FOR loop in our Roman Numeral conversion program. Line 69 contains the *control section*. Like the example above, the first expression, I = 0, initializes the *control variable* to 0. However, the *conditional expression*, I < INPUTSTRING.LENGTH(), is not as straightforward.

To ensure that it's converted every character entered by the user, the loop must process every Roman letter in the string, INPUTSTRING. To accomplish this, the loop increments I (via the *increment expression*, I++) and compares its value to the string length of INPUTSTRING (I < INPUTSTRING.LENGTH()) during each iteration. In simpler terms, I "steps through" every letter in INPUTSTRING. (This will become clear shortly.)

Now let's describe the code contained in the loop's *body* that executes during each iteration. We can ignore the comments that appear on lines 71–73; the first executable statement appears on line 74:

```
ROMANVALUE = CONVERTROMANCHARTOINTEGERVALUE( INPUTSTRING.CHARAT(I) );
```

Several operations take place within this one statement. First, the expression, INPUTSTRING. CHARAT(I), extracts the Roman letter at the current processing location. For example, let's assume the user entered the Roman Numeral "LXV". In INPUTSTRING, "L" is in position 0, "X" is in position 1, and "V" is in position "2". (Sorry to harp, but remember we begin counting at zero.) The first time through the loop, when I equals 0, the expression INPUT-STRING.CHARAT(0) yields the value "L". The second time through the loop, I equals 1, and the expression INPUTSTRING.CHARAT(1) evaluates to "X"; the last time through the loop, I equals 2, and the expression INPUTSTRING.CHARAT(2) results in the value "V".

Next, the value of the expression INPUTSTRING.CHARAT(I) is passed as a parameter to the method CONVERTROMANCHARTOINTEGERVALUE(). As we saw in the prior section, this method returns the integer value of the Roman letter it receives.

Finally, we store the value returned by CONVERTROMANCHARTOINTEGERVALUE() in the variable ROMANVALUE.

To summarize, each time this statement executes, ROMANVALUE contains the equivalent integer value of the Roman letter positioned at location I in INPUTSTRING.

The method's next task is to determine whether the letter it's currently processing is a single character value or it's part of a *two-character sequence* (see Rule 3 above). The first test appears on line 76. This is an example of a Java *if statement*. It looks like a *for loop*, but there is no iteration.

We interpret an IF statement as follows: if the *conditional expression* enclosed within the parentheses yields TRUE, then execute code contained in the *body* (i.e., all the statements residing between the curly braces—lines 78–108 in the listing). Otherwise, if the *conditional expression* yields FALSE, jump to the code contained in the ELSE block (lines 109–115 in the listing).

Let's review this in more detail.

The expression on line 76, I < INPUTSTRING.LENGTH()-1, asks the question: *Are there more characters after this one?* It does this by determining whether the current value of I is equal to the length of the INPUTSTRING. If there are more characters (i.e., I's value is less than the string length of INPUTSTRING), the result is TRUE. Otherwise, the result is FALSE.

Let's start with the simple case. If the result is FALSE, this is indeed the final character in the Roman string. Therefore, by definition, there cannot be any characters to its right, and we can be certain that the current character is NOT part of a *two-character sequence*. Thus, the algorithm executes the statement appearing on line 114 (INTVALUE += ROMANVALUE;), which adds the current character's value to the running total stored in INTVALUE.

We interpret "+=" operator as follows:

- Get the value currently stored in INTVALUE (the running total)

- Add to it the value stored in ROMANVALUE (the value of the current Roman letter)

- Store the computed sum back in INTVALUE (update the running total)

The more complicated case is when the method is processing a character other than the last one. If the *conditional statement* on line 76 evaluates to TRUE, we know that there is at least one Roman letter to the right of the one we're currently processing. Thus, the current character *might* be part of a *two-character sequence.*

To determine this, we need to ascertain whether the next character in sequence is greater in value than the current one. If it is, the two characters form a *two-character sequence,* and we must subtract the two values (per Rule 3) and add the result to the running total. Otherwise, we add the current letter's value to our running total and ignore the next character (for the moment).

Let's look at how the code does this.

The first step of this processing appears on line 82 when the algorithm computes the value of the *next* Roman letter (i.e., the letter to the right of the current one). Following that, on line 84, the code compares the two values: if the current letter is greater than or equal to the letter to its right, we add the current letter's value to our running total (line 92).

However, if the *next* letter's value is greater than that of the current letter, it's part of a *two-character sequence*, and the code must subtract the current letter's value from the *next* letter's value then add that result to the running total.

The code to perform this computation appears on line 101:

```
INTVALUE += (NEXTROMANVALUE - ROMANVALUE);
```

As with math formulas, any Java expression enclosed between parentheses takes precedence. Thus, the subtraction (NEXTROMANVALUE - ROMANVALUE) executes first. Then, as indicated by the "+=" operator, the code adds the result of the subtraction to the running total contained in INTVALUE.

There's one more bit of housekeeping that we must address. That last computation required two Roman letters: the *current* one and the one to its right (i.e., the *next* one). However, based on the *increment expression* (I++) in the enclosing FOR loop (line 69 in the listing), the algorithm will erroneously process that same *next* character during its next iteration. Unless we take some action, the loop will incorrectly process the *next* letter twice.

To account for *two-character sequences*, we need to adjust the counter (I) to point to the *next* letter, not the current one. Then, when the loop *increment expression* executes, the counter will point to the first unprocessed letter in INPUTSTRING. The statement appearing on line 107 (I++;) addresses this issue.

That completes our review of the conversion program. Again, please don't feel concerned if you didn't absorb every nuance of the code during your first reading. Like anything worthwhile in life, it takes time. Programmers study for years to hone their craft. Nonetheless, I hope you gained some insight into the design and coding of software applications.

The next section demonstrates how to bring this program to life.

Compile and Execute

As you may recall from Chapter 10, there are two main ways programs execute: compilation and interpretation. Java is a hybrid language in that it uses a compiler to translate source code into a pseudo-machine language called *bytecode*. Once compiled, a utility called the Java Virtual Machine (JVM) interprets and executes the compiled bytecode.

> *Bytecode* is a machine-like instruction set optimized for execution by a software interpreter or virtual machine.

In most environments, the name of the Java compiler is JAVAC. We invoke it as follows:

JAVAC [FILENAME.JAVA]

As noted above, I chose to name the Roman-to-integer conversion program RomanToInteger. As a result, Java requires that the source code resides in a file named RomanToInteger.JAVA. Hence, the actual command line to compile our example is as follows:

JAVAC RomanToInteger.JAVA[26]

If there are no errors in the source code, the compiler generates and stores the *bytecode* in a file called RomanToInteger.CLASS. However, it's rare (dare I say unheard of) to write a program that runs the first time without any errors. On the contrary, developers routinely engage in a repeated process of *editing, compiling, and testing* until the program functions properly. [27]

To make the process appear a bit more realistic, I deliberately introduced[28] some errors into the source code appearing in Listing 12.1 so that you see how the compiler responds to them. I've noted the errors below.

On line 74, I left off the "e" on the variable RomanValue, as follows:

RomanValu = convertRomanCharToIntegerValue(inputString.charAt(i));

On line 92, I omitted the semicolon:

intValue += RomanValue

And, on line 101, I excluded the right paren:

intValue += (nextRomanValue - RomanValue;

The screenshot depicted in Figure 12.1 is from a program called the BASH shell.[29] Like Window's COMMAND PROMPT, it's one of several command-line interfaces provided by the Cygwin platform.[30] The string, ENTER COMMAND:,[31] is a prompt issued by the BASH shell to alert the user (me in this case) that it's ready to execute the next command. (The prompt in a Window's environment might look something like this: "c:\>").

---

[26] Note that, in this case, case matters.
[27] When developing this program, I executed the compiler many times and had to fix numerous errors.
[28] Anyway, that's my story.
[29] The designation, *shell*, is a generic term that describes any command-line interpreter; BASH is but one of many available to programmers.
[30] I ran all these examples under Cygwin; it's my preferred development environment. However, everything presented in this section should execute identically using COMMAND PROMPT under Windows.
[31] The prompt string is customizable; I chose "Enter Command:" for clarity.

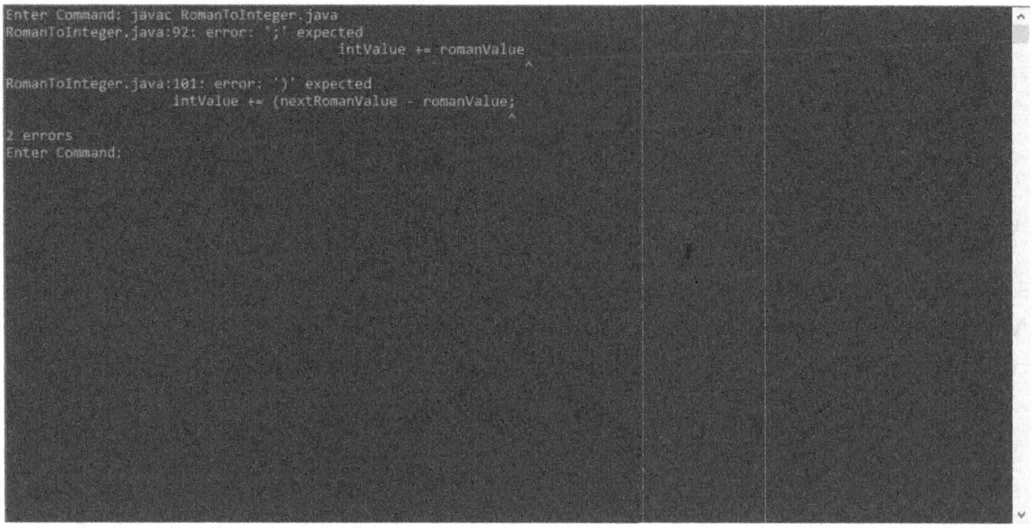

FIGURE 12.1 JAVAC: 1st Invocation.

At the top of the display, you can see where I executed the Java compiler, JAVAC,[32] for the first time; it identified two of the three errors and generated diagnostic messages for lines 92 and 101. It's often the case that some errors might prevent the compiler from uncovering other errors when it executes.

I corrected the two mistakes and recompiled the program.

As highlighted in Figure 12.2, the compiler identified the third error this time. I corrected that issue and recompiled the program one more time.

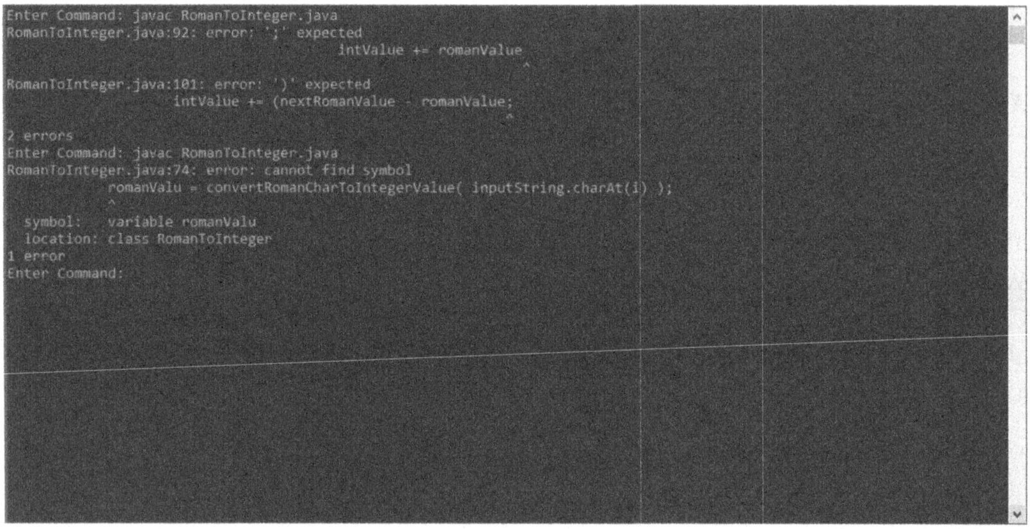

FIGURE 12.2 JAVAC: 2nd Invocation

---

[32] The command names and their corresponding output should be identical in a Windows environment.

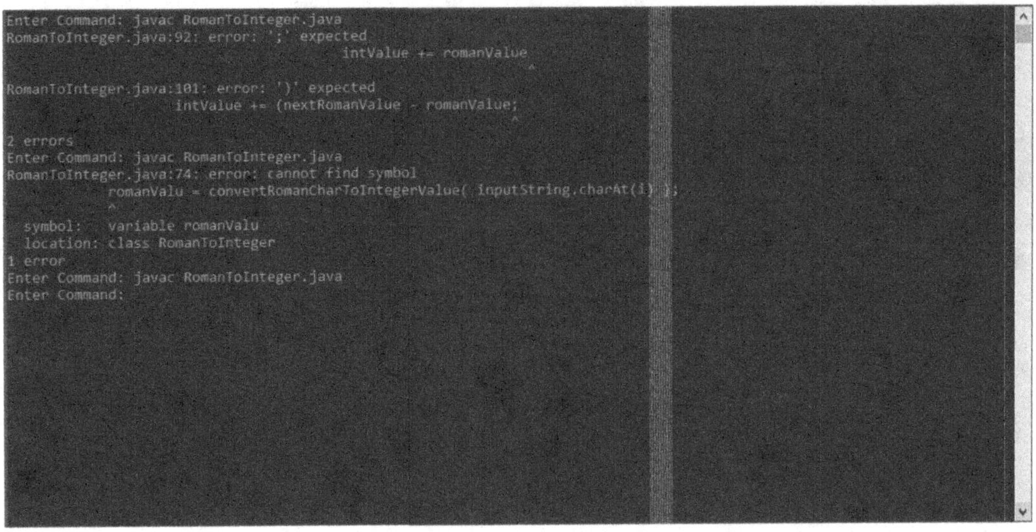

FIGURE 12.3   JAVAC: 3rd Invocation—Success!

Voilà—the third time's the charm (see Figure 12.3). Note that when there are no errors, there is no output from the compiler. (As a rule in the IT world, no news is good news.)

As depicted in Figure 12.4, when successful, the compiler creates a file called RomanToInteger.class, which contains the executable *bytecode* for our program. To display the files in my folder, I used a command called ls.[33] The asterisk ("*") serves as a wildcard character that matches any value. That's why both the RomanToInteger.class and the RomanToInteger.java files appeared in the output.

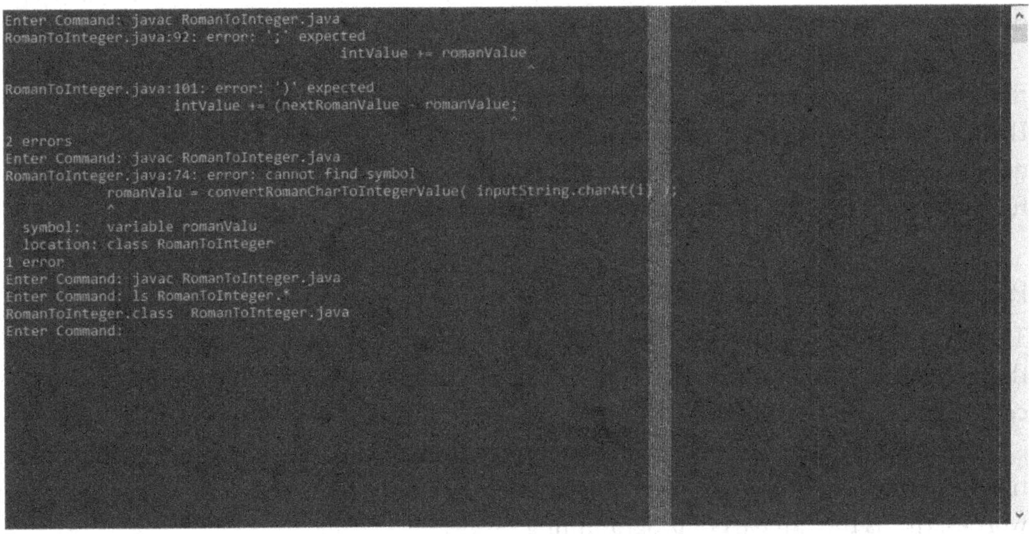

FIGURE 12.4   Compilation result.

---

[33] This is equivalent to the DIR command under Windows.

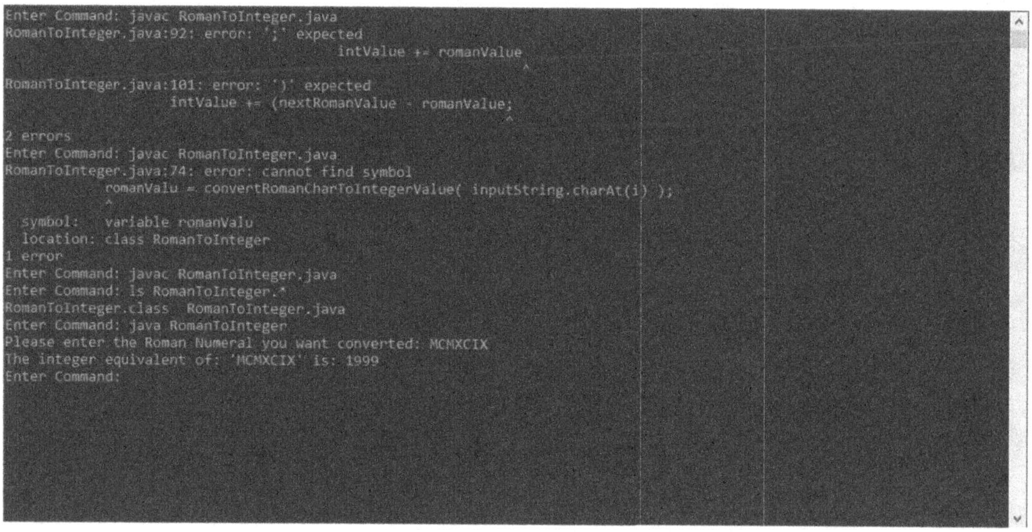

FIGURE 12.5 RomanToInteger —1st Execution.

To execute our program, we must invoke the JVM and indicate which file to execute. To do that, I entered the following command:

JAVA RomanToInteger

Note that I didn't specify an extension; the JAVA program assumes that the file name I provided ends with .CLASS.

Figure 12.5 displays the result of the first execution. At the prompt, I entered MCMXCIX, and the program computed and displayed the correct result: 1999.

One last point. There are many other tools I could have used to compile and execute this program. Today, most Java developers use Integrated Development Environments (or IDEs),[34] which incorporate editors, compilers, and debuggers into a single development dashboard. Although they vary in the features they provide, they all are extremely powerful. However, I thought them overkill for this example.

PART 2—INTEGER TO ROMAN

The next program we'll write performs the compliment operation: it converts integers into their equivalent Roman Notation. For example, given a value of 1999, the program will compute the result as "MCMXCIX."

Unfortunately, however, converting integers to Roman Notation is a complement operation in name only. That is, to solve this problem, we must alter our design approach. No worries: this is the fun part of programming.

---

[34] Many IDEs are free to download and use.

To get started, consider the integer value, 2999. When we examine the leftmost digit (i.e., the "2" in the thousand's column), we immediately think "MM." That mental exercise is so simple and straightforward that it might leave us with the (erroneous) impression that all we need to do is successively divide the integer quantity by the value of each Roman letter to compute the resulting Roman Numeral.

However, as soon as we move to the next digit, "9", it becomes evident that the solution is not that simple: representing the value 900 in Roman Numerals requires a *two-character sequence*, "CM."

Clearly, we must alter our approach. As the last example demonstrates, whatever solution we develop must consider values beyond the basic set of Roman letters. It must also address integer values that require *two-character* sequences (e.g., CM, CD, XC, etc.).

However, there's one property of *two-character sequences* that we can use to our advantage: each may appear only once (at most) in an integer value. For example, you can't have multiple values of "90"—"XC"—in an integer. If there were two of them, the integer value would be "180" (2 * 90), which in Roman notation is "CLXXX", not "XCXC". (The latter construct is invalid.) We'll leverage that knowledge as we design the solution.

Requirements

The requirements for our Integer-to-Roman conversion are as follows:

1. Upon invocation, the program will prompt the user to enter an integer value.

2. After the user enters the value, the program displays the equivalent Roman notation.

3. The user may enter any valid integer between the values 1 and 3999 (as permitted in basic Roman notation).

4. As per our first coding example, the program will support a simple line-oriented user interface.

Analysis

As described above, the program must calculate the type and quantity of Roman letters required to represent the integer value it is to convert. Thus, the code must function as follows:

- For every 1,000 in value (at most 3), display an "M"
- Display "CM" if the integer contains the value 900
- For every 500 in value (there should be only one), display a "D"
- Display "CD" if the integer contains the value 400
- For every 100 in value (at most 3), display a "C"

The program continues in this manner for all remaining valid one- and two-letter Roman Numerals until it completes the conversion.

Design

Overall, the design of this program is like that of our first example.

- Set up the environment—declare and initialize all variables

- Prompt the user for input using a built-in Java function.

- Read and store the integer value entered by the user (using another built-in function).

- Compute and display the equivalent Roman notation.

Code

Program Listing

The code for the Integer-to-Roman conversion program appears in Listing 12.5. We'll discuss the solution in detail in the sections that follow.

LISTING 12.5 INTEGER TO ROMAN CONVERSION

```
001 //
002 // CONVERT INTEGER VALUES TO ROMAN NOTATION
003 //
004
005 IMPORT JAVA.UTIL.SCANNER;                              // REQUIRED TO USE JAVA'S
                                                              SCANNER LIBRARY
006
007 //
008 // PROGRAM/CLASS NAME
009 //
010 PUBLIC CLASS IntegerToRoman
011 {
012     STATIC FINAL INT MIN_ROMAN_VALUE = 1;              // DEFINE LITERALS
013     STATIC FINAL INT MAX_ROMAN_VALUE = 3999;
014
015     //
016     //      CONVERT INTEGER VALUES TO ROMAN NOTATION
017     //
018     //
019     PUBLIC STATIC String convertIntegerToRomanNumeral ( INT INPUTVAL )
020     {
021             String romanString = "";   // VARIABLE TO STORE NUMERALS
022
023             WHILE ( INPUTVAL >= 1000 ) // COUNT 1,000's -> M
024             {
025                     romanString += "M";
026                     INPUTVAL -= 1000;
027             }
028
```

```
029             IF( INPUTVAL >= 900 )              // 900? -> CM
030             {
031                     ROMANSTRING += "CM";
032                     INPUTVAL -= 900;
033             }
034
035             WHILE( INPUTVAL >= 500 )           // Count 500's -> D
036             {
037                     ROMANSTRING += "D";
038                     INPUTVAL -= 500;
039             }
040
041             IF( INPUTVAL >= 400 )              // 400? -> CD
042             {
043                     ROMANSTRING += "CD";
044                     INPUTVAL -= 400;
045             }
046
047             WHILE( INPUTVAL >= 100 )           // Count 100's -> C
048             {
049                     ROMANSTRING += "C";
050                     INPUTVAL -= 100;
051             }
052
053             IF( INPUTVAL >= 90 )               // 90? -> XC
054             {
055                     ROMANSTRING += "XC";
056                     INPUTVAL -= 90;
057             }
058
059             WHILE( INPUTVAL >= 50 )            // Count 50's -> L
060             {
061                     ROMANSTRING += "L";
062                     INPUTVAL -= 50;
063             }
064
065             IF( INPUTVAL >= 40 )               // 40? -> XL
066             {
067                     ROMANSTRING += "XL";
068                     INPUTVAL -= 40;
069             }
070
071             WHILE( INPUTVAL >= 10 )            // Count 10's -> x
072             {
073                     ROMANSTRING += "X";
074                     INPUTVAL -= 10;
075             }
076
077             IF( INPUTVAL >= 9 )                // 9? -> Ix
078             {
079                     ROMANSTRING += "IX";
080                     INPUTVAL -= 9;
081             }
082
```

```
083            WHILE ( INPUTVAL >= 5 )              // COUNT 5's -> V
084            {
085                    ROMANSTRING += "V";
086                    INPUTVAL -= 5;
087            }
088
089            IF ( INPUTVAL >= 4 )                 // 4? -> IV
090            {
091                    ROMANSTRING += "IV";
092                    INPUTVAL -= 4;
093            }
094
095            WHILE ( INPUTVAL >= 1 )              // COUNT 1's -> I
096            {
097                    ROMANSTRING += "I";
098                    INPUTVAL -= 1;
099            }
100
101            //
102            //      RETURN THE GENERATED STRING
103            //
104            RETURN ( ROMANSTRING );
105    }
106
107    //
108    //      PROCESSING BEGINS HERE
109    //
110    PUBLIC STATIC VOID MAIN ( STRING [] ARGS )
111    {
112            //
113            //      DECLARE VARIABLES ("STORAGE LOCKERS")
114            //
115            INT     ORIGINTVALUE;
116            SCANNER IN;
117            STRING  ORIGSTRING;
118            STRING  ROMANSTRING;
119
120            //
121            //      THE LINE OF CODE PREPARES THE SYSTEM
122            //      TO ACCEPT INPUT
123            //
124            IN = NEW SCANNER ( SYSTEM.IN );
125
126            //
127            //      PROMPT THE USER TO TYPE IN THE STRING
128            //
129        SYSTEM.OUT.PRINT ( "PLEASE ENTER THE INTEGER YOU WANT CONVERTED: " );
130
131            //
132            //      READ WHAT THE USER TYPED
133            //      STORE IT IN 'ORIGSTRING'
134            //
135        ORIGSTRING = IN.NEXTLINE ();
```

```
136
137         //
138         //        CONVERT THE INPUT STRING INTO AN INTEGER VALUE
139         //
140      origIntValue = Integer.parseInt( origString );
141
142         //
143         //        DETERMINE IF INPUT IS WITHIN RANGE
144         //
145      if( origIntValue < MIN_ROMAN_VALUE || origIntValue > MAX_ROMAN_
         VALUE )
146      {
147             //
148             //        INCORRECT INPUT -> INFORM USER
149             //
150             System.out.println( "Invalid Roman Number Value: '" + origString
                + "'" );
151             System.out.println( "Program exiting!" );
152             System.exit( 0 );
153      }
154
155         //
156         //        CONVERT THE INTEGER VALUE TO A Roman Numeral
157         //
158      romanString = convertIntegerToRomanNumeral( origIntValue );
159
160         //
161         //        DISPLAY RESULT TO THE USER
162         //
163      System.out.println( "Roman equivalent of: '" + origString + "' is: '" +
         romanString + "'" );
164   }
165 }
```

The MAIN() Method

As you review Listing 12.5, you'll notice that much of the code looks like our previous program:

- The first few lines contain some comments.

- The IMPORT statement (line 5) includes the code for the SCANNER class.

- Line 10 contains the CLASS declaration—although, in this case, we've named the program INTEGERTOROMAN. (That means the code must reside in a file named INTEGERTOROMAN.JAVA.)

However, lines 12 and 13 contain constructs that we haven't seen before. These two statements define *symbolic constants* representing the minimum (MIN_ROMAN_VALUE) and maximum (MAX_ROMAN_VALUE) integer values the program will convert.

Programmers use *symbolic constants* for several reasons:[35]

- They are mnemonic and aid *readability*. For example, if we saw a literal value of "1" used in a line of code, we might not understand what it represents. However, when we read MIN_ROMAN_VALUE, we have no doubt.

- If we must change a *symbolic constant*'s value for some reason, we need to edit only one line of code. (Line 12 or 13 in our example.) Thus, we don't have to scour the code searching for all instances of the literal and making repetitive edits.

As with our first program, execution begins with the MAIN() method (line 110). So, let's start there; we'll return to the CONVERTINTEGERTOROMANNUMERAL() method in the next section. The first few lines of code in MAIN() should look familiar; it:

- Declares variables (lines 115 through 118);

- Instantiates a SCANNER object (line 124);

- Prompts the user to enter a value (line 129); and

- Reads the user's response (line 135).

Beginning with line 140, we see some new code. As you may recall from Chapter 3, computer keyboards transmit each keypress as an ASCII character value. This includes digits.

For example, if you type "1234<ENTER>", the keyboard sends five distinct ASCII codes to the computer. (One for each digit and one for the ENTER key.) Even though we intuitively think of "1234" as a single four-digit numeric quantity, the four characters comprising it are individual keypresses in the world of computers.

Thus, before it can process a series of individual digits as a single quantity, the program needs to convert them into a numeric value—an integer in this case. The code appearing on line 140 performs that function.

```
ORIGINTVALUE = INTEGER.PARSEINT( ORIGSTRING );
```

The built-in Java method, INTEGER.PARSEINT, accepts a STRING (ORIGSTRING in this case) and converts it into an integer value. The code stores the result of this conversion in the variable ORIGINTVALUE.

Following the integer conversion, the code appearing on lines 145 through 153 validates the input it received. Specifically, the program ensures that the user entered a number between 1 and 3999. If not, it displays two error messages (lines 150 and 151) and then exits (line 152).

If the user entered a valid number, the conversion begins on line 158 with the following statement:

```
ROMANSTRING = CONVERTINTEGERTOROMANNUMERAL( ORIGINTVALUE );
```

---

[35] There are other reasons, but they are beyond the scope of this text.

We'll discuss how this method works in the next section.

When CONVERTINTEGERTOROMANNUMERAL() returns, MAIN() displays the result using the PRINTLN() method (line 163).

The CONVERTINTEGERTOROMANNUMERAL() Method

Beginning at line 19 of the listing, the CONVERTINTEGERTOROMANNUMERAL() method contains the code that performs the conversion. It accepts one parameter—the integer value the user would like transformed into Roman Notation. Line 21 declares the variable, ROMANSTRING, which will contain the result of the conversion. The actual conversion code begins with line 23.

Let's see how it works.

As per our analysis (see above), the program initiates each conversion by determining how many "thousands" are in the integer value. For example, 3999 has three—and its corresponding Roman Notation would begin with three Ms (e.g., "MMM").

That result is so obvious as to be trivial. However, that's not the case with a computer program. Unlike humans, the code can't just *look* at an integer and *know* how many thousands it contains. Instead, it must somehow *compute* the answer. The question is: how?

One way to do it is as follows:

- Step 1: The program tests whether the integer value is greater than 1,000.

- Step 2: If it is:

   a. Step 2a: Count 1 "M"

   b. Step 2b: Subtract 1,000 from the original integer value

   c. Step 2c: Repeat Step 1

- Step 3: If not (i.e., the integer value is less than 1,000), move on to the next Roman letter

The code contained on lines 23–27 implements the approach just described using a WHILE loop. However, before explaining the solution in detail, we need to discuss how Java WHILE loops work—they are similar to FOR loops.

As depicted in the code template below, a Java WHILE statement comprises two parts: a *test condition* (positioned between parentheses) and a *statement block* or *body* (contained within the curly braces).

```
WHILE ( TEST CONDITION)
{
        STATEMENT 1;
        STATEMENT 2;
        ...
        STATEMENT N;
}
```

At the start of each iteration (including the first), the *test condition* executes. If it evaluates to TRUE, each statement[36] in the *body* of the loop executes in sequence. Whenever the *test condition* evaluates to FALSE, the loop terminates, and execution resumes with the statement immediately following the closing brace. Note that if the *test condition* evaluates to FALSE during the first iteration, the WHILE loop's *body* will not execute at all.

Now, returning to our program, the expression appearing on line 23 of the listing,

```
WHILE ( INPUTVAL >= 1000 )
```

asks the question: Is the value of INPUTVAL *greater than* or *equal* to 1,000? If it is, the *body* of the loop executes:

- Line 25 appends an "M" to the variable, ROMANSTRING, using the "+=" operator. (This is how the algorithm constructs the Roman Numeral—one symbol at a time.)
- Line 26 subtracts 1,000 from INPUTVAL. This step ensures that we generate the correct number of Ms in the output.

The iteration continues in this manner until INPUTVAL is less than 1,000. When that happens, the loop has generated the appropriate number of Ms. After the WHILE loop terminates, the method moves on to the next step in the conversion (as we'll see shortly).

Note that this code functions correctly even for integer values less than 1,000. For example, if the user entered the value 253, INPUTVAL is less than 1,000 during the first iteration of the WHILE loop. In this case, the *test condition* would yield FALSE the first time, and the loop's *body* (lines 25 and 26) won't execute at all, and the resulting output will contain no Ms.

Please take a moment to review this approach. It serves as a model for the entire conversion process that follows.

After processing the Ms, the algorithm must determine whether INPUTVAL is greater than or equal to 900. If it is, the code must append the *two-character sequence* "CM" to ROMANSTRING. That processing appears on lines 29–33.

Although similar to the approach used to process Ms, there is a slight modification to the code. As mentioned above, *two-character sequences* may appear only once in a Roman Numeral. Thus, there is no need to loop, so we replaced the WHILE loop with an IF statement. (A Java IF statement works like a WHILE loop except that it will execute its statement *body* only once if its *test condition* evaluates to TRUE; there's no implied iteration.)

The method continues in this manner—alternating WHILE loops and IF statements—until it completes the conversion. On line 104, it returns the converted Roman Numeral to its caller, MAIN().

Compile and Execute

Except for a change in file name (as noted above), compiling and executing this program is identical to our first example. The screenshot appearing in Figure 12.6 captures the process and demonstrates some conversions.

---

[36] This can be any valid Java construct including nested WHILE loops.

```
Enter Command:
Enter Command:
Enter Command: javac IntegerToRoman.java
Enter Command: java IntegerToRoman
Please enter the Integer you want converted: 3999
The Roman Numeral equivalent of: '3999' is: 'MMMCMXCIX'
Enter Command: java IntegerToRoman
Please enter the Integer you want converted: 2020
The Roman Numeral equivalent of: '2020' is: 'MMXX'
Eqter Command: java IntegerToRoman
Please enter the Integer you want converted: 601
The Roman Numeral equivalent of: '601' is: 'DCI'
Enter Command:
```

FIGURE 12.6   Integer to Roman Conversion.

ADVANCED TOPIC: COMBINED PROGRAM

We now have two working solutions: one converts integers to Roman Notation; the other performs the compliment operation. However, as designed, they are two separate, independent executables. Thus, users must decide which one to invoke to accomplish the desired conversion. Moreover, the two programs share a lot of duplicate/similar code.

Though not the end of the world, this approach is not ideal—and we can certainly do better. Thus, being the competent, conscientious coders we are, let's combine both programs into a single executable.

The next section discusses the revised design.

Modified Design

To combine the solutions, the first point we must consider is how to merge the code. Unfortunately, if we mechanically copy both programs into a new file, we'd create two issues.

First, the new file would contain multiple CLASS names (i.e., RomanToInteger and IntegerToRoman—lines 7 and 10, respectively, in the two original program listings). This approach would make it difficult to name the new file. (Please recall that Java determines file names based on CLASS names. Having two classes in one file would create a naming conflict.)[37]

What's worse, the combined version would have two MAIN() methods.[38] The duplication would make it challenging for the JVM to determine where program execution should begin.

---

[37] Though not recommended, Java does permit multiple classes in one source file (with certain constraints). However, the details are beyond the scope of this text.

[38] Strictly speaking, Java permits two MAIN() methods in one program. The details, however, are beyond the scope of this text.

The first issue is the easier one to address. We'll create a new CLASS called COMBINEDCONVERSION and, except for the two versions of MAIN(), it will include every method contained in both programs.

Conceptually, the second issue is just as easily solved: we'll include only one version of MAIN()in the new program. However, there are some practical details we must address. Specifically, how will the new MAIN() determine which conversion to execute (i.e., ROMANTOINTEGER or INTEGERTOROMAN) and, by extension, select which method to invoke?

The direct approach is to ask the user. The program could issue a prompt that might look something like this:

TYPE 1 FOR ROMAN TO INTEGER OR 2 FOR INTEGER TO ROMAN:

Based on the user's response (either "1" or "2"), the program could determine which conversion to invoke. Of course, following that, the program would still have to prompt the user to enter a value to convert. Although we now have only one program, the user must respond to two prompts. This approach moves us in the right direction, but I think we can do even better.

Let's consider the program's input for a moment. When converting a Roman Numeral, the user enters *letters*. When converting an integer, the user enters *digits. Voilà!* The values are *orthogonal.* That is, they're either letters *or* digits, never both. Thus, the program can determine the type of conversion as follows:

- If the value entered by the user contains only *letters*, execute ROMANTOINTEGER

- If the value entered by the user comprises only *digits*, invoke INTEGERTOROMAN

We'll present the code for the combined solution in the next section.

Program Listing

Listing 12.6 contains the source code for the file COMBINEDVERSION.JAVA, the merged version of the conversion programs.

LISTING 12.6 MERGED CONVERSION PROGRAMS

```
001 //
002 // COMBINED PROGRAM -
003 //        CONVERT INT -> ROMAN AND ROMAN -> INT
004 //
005
006 IMPORT JAVA.UTIL.SCANNER;              // REQUIRED TO USE JAVA'S SCANNER LIBRARY
007
008 //
009 // PROGRAM/CLASS NAME
010 //
011 PUBLIC CLASS COMBINEDCONVERSION
```

```
012  {
013       STATIC FINAL INT MIN_ROMAN_VALUE = 1;
014       STATIC FINAL INT MAX_ROMAN_VALUE = 3999;
015
016       //      ========= CODE TO CONVERT ROMAN TO INTEGER
017
018       //
019       //      THIS FUNCTION WILL CONVERT A ROMAN
020       //      CHARACTER INTO ITS EQUIVALENT INTEGER VALUE
021       //
022       PUBLIC STATIC INT CONVERTROMANCHARTOINTEGERVALUE( CHAR ROMANCHAR )
023       {
024               //
025               //      DETERMINE THE CHARACTER AND
026               //      RETURN THE APPROPRIATE VALUE
027               //
028               SWITCH( ROMANCHAR )
029               {
030                       CASE 'I':
031                           RETURN( 1 );
032
033                       CASE 'V':
034                           RETURN( 5 );
035
036                       CASE 'X':
037                           RETURN( 10 );
038
039                       CASE 'L':
040                           RETURN( 50 );
041
042                       CASE 'C':
043                           RETURN( 100 );
044
045                       CASE 'D':
046                           RETURN( 500 );
047
048                       CASE 'M':
049                           RETURN( 1000 );
050
051                       DEFAULT:
052                           //
053                           // BAD CHARACTER - FLAG AS ERROR
054                           // INFORM THE USER
055                           //
056                           SYSTEM.OUT.PRINTLN( "INVALID ROMAN CHARACTER: '" +
                                   ROMANCHAR + "'" );
057                           RETURN( 0 );
058               }
059       }
060
061       //
062       //      THIS FUNCTION CONVERTS ROMAN NUMERALS TO INTEGERS
063       //      THIS IS WHERE THE 'HEAVY LIFTING' TAKES PLACE
064       //
```

```
065    PUBLIC STATIC INT convertRomanStringToInt( String inputString )
066    {
067            //
068            //      Declare variables ("storage lockers")
069            //
070            INT i = 0;
071            INT intValue = 0;
072            INT romanValue = 0;
073            INT nextRomanValue = 0;
074
075            //
076            // Step through every character in the string
077            //
078            FOR( i = 0; i < inputString.length(); i++ )
079            {
080                    //
081                    // Get the value of the current character
082                    //
083        romanValue = convertRomanCharToIntegerValue( inputString.charAt(i) );
084
085                    IF( i < inputString.length()-1 )   // Are there more characters?
086                    {
087                            //
088                            //      There are more characters -
089                            //      Get the integer value of the 'next' character
090                            //
091                    nextRomanValue = convertRomanCharToIntegerValue( inputString.
                       charAt(i+1) );
092
093                            IF( RomanValue >= nextRomanValue )
094                            {
095                                    //
096                                    //      This is the 'normal' case
097                                    //      e.g., 'vi' or 'lx'
098                                    //      We can ignore the 'next' value and
099                                    //      add the current value to the running
                                          total
100                                    //
101                                    intValue += romanValue;
102                            } ELSE IF( romanValue < nextRomanValue ) {
103                                    //
104                                    //      This is the case when we have to
                                          subtract values
105                                    //      e.g., 'iv' or 'xl'
106                                    //      In this case we have to subtract the
                                          'current' value
107                                    //      from the 'next' value and add that
                                          result
108                                    //      to the running total.
109                                    //
110                                    intValue += (nextRomanValue - romanValue);
111
112                                            //
```

```
113                                    //        WE ALSO HAVE TO 'CONSUME' THE 'NEXT'
                                                 VALUE
114                                    //        BECAUSE WE'VE ALREADY PROCESSED IT.
115                                    //
116                    I++;
117                  }
118              } ELSE {
119                            //
120                            //    NO MORE CHARACTERS AFTER THIS ONE ...
121                            //    ADD THE LAST CHARACTER TO RUNNING TOTAL
122                            //
123                            INTVALUE += ROMANVALUE;
124                  }
125          }
126
127          //
128          //    RETURN THE RESULT
129          //
130          RETURN( INTVALUE );
131    }
132
133    //    ========= CODE TO CONVERT INTEGER TO ROMAN
134
135    //
136    //    CONVERT INTEGER VALUES TO ROMAN NOTATION
137    //
138    PUBLIC STATIC STRING CONVERTINTEGERTOROMANNUMERAL( INT INPUTVAL )
139    {
140          STRING ROMANSTRING = "";  // VARIABLE TO STORE NUMERALS
141
142          WHILE( INPUTVAL >= 1000 )// COUNT 1,000's -> M
143          {
144                  ROMANSTRING += "M";
145                  INPUTVAL -= 1000;
146          }
147
148          IF( INPUTVAL >= 900 )              // 900? -> CM
149          {
150                  ROMANSTRING += "CM";
151                  INPUTVAL -= 900;
152          }
153
154          WHILE( INPUTVAL >= 500 ) // COUNT 500's -> D
155          {
156                  ROMANSTRING += "D";
157                  INPUTVAL -= 500;
158          }
159
160          IF( INPUTVAL >= 400 )              // 400? -> CD
161          {
162                  ROMANSTRING += "CD";
163                  INPUTVAL -= 400;
164          }
```

```
165
166            WHILE( INPUTVAL >= 100 ) // COUNT 100's -> C
167            {
168                    ROMANSTRING += "C";
169                    INPUTVAL -= 100;
170            }
171
172            IF( INPUTVAL >= 90 )              // 90? -> XC
173            {
174                    ROMANSTRING += "XC";
175                    INPUTVAL -= 90;
176            }
177
178            WHILE( INPUTVAL >= 50 )          // COUNT 50's -> L
179            {
180                    ROMANSTRING += "L";
181                    INPUTVAL -= 50;
182            }
183
184            IF( INPUTVAL >= 40 )              // 40? -> XL
185            {
186                    ROMANSTRING += "XL";
187                    INPUTVAL -= 40;
188            }
189
190            WHILE( INPUTVAL >= 10 )          // COUNT 10's -> X
191            {
192                    ROMANSTRING += "X";
193                    INPUTVAL -= 10;
194            }
195
196            IF( INPUTVAL >= 9 )               // 9? -> IX
197            {
198                    ROMANSTRING += "IX";
199                    INPUTVAL -= 9;
200            }
201
202            WHILE( INPUTVAL >= 5 )           // COUNT 5's -> V
203            {
204                    ROMANSTRING += "V";
205                    INPUTVAL -= 5;
206            }
207
208            IF( INPUTVAL >= 4 )               // 4? -> IV
209            {
210                    ROMANSTRING += "IV";
211                    INPUTVAL -= 4;
212            }
213
214            WHILE( INPUTVAL >= 1 )           // COUNT 1's -> I
215            {
216                    ROMANSTRING += "I";
217                    INPUTVAL -= 1;
218            }
```

```
219
220          //
221          //        RETURN THE GENERATED STRING
222          //
223          RETURN( ROMANSTRING );
224     }
225
226  //
227  //     PROGRAM EXECUTION BEGINS HERE
228  //
229  PUBLIC STATIC VOID MAIN( STRING[] ARGS )
230  {
231          //
232          // DECLARE VARIABLES
233          //
234          INT     INTVAL;
235          INT     ORIGINTVALUE;
236          SCANNER IN;
237          STRING  ORIGSTRING;
238          STRING  ROMANSTRING;
239
240          //
241          //        PROMPT USER AND READ INPUT
242          //
243          IN = NEW SCANNER( SYSTEM.IN );
244      SYSTEM.OUT.PRINT( "PLEASE ENTER THE VALUE YOU WANT CONVERTED: " );
245      ORIGSTRING = IN.NEXTLINE();
246
247          //
248          // TEST TO SEE WHAT TYPE OF CONVERSION THE USER WANTS
249          //
250          IF( CHARACTER.ISDIGIT(ORIGSTRING.CHARAT(0)) )
251          {
252                  //
253                  // CONVERT FROM INTEGER TO ROMAN
254                  //
255                  ORIGINTVALUE = INTEGER.PARSEINT( ORIGSTRING );
256                  IF( ORIGINTVALUE < MIN_ROMAN_VALUE || ORIGINTVALUE >
                    MAX_ROMAN_VALUE )
257                  {
258                          SYSTEM.OUT.PRINTLN( "INVALID ROMAN NUMBER VALUE: '" +
                            ORIGSTRING + "'" );
259                          SYSTEM.OUT.PRINTLN( "PROGRAM EXITING!" );
260                          SYSTEM.EXIT( 0 );
261                  }
262
263                  ROMANSTRING = CONVERTINTEGERTOROMANNUMERAL( ORIGINTVALUE );
264
265                  SYSTEM.OUT.PRINTLN( "THE ROMAN EQUIVALENT OF: '" + ORIGSTRING +
                    "' IS: '" + ROMANSTRING + "'" );
266          } ELSE {
267                  //
268                  // CONVERT FROM ROMAN TO INTEGER
269                  //
```

```
270                          INTVAL = CONVERTROMANSTRINGTOINT ( ORIGSTRING . TOLOWERCASE () ) ;
271                          SYSTEM . OUT . PRINTLN ( "THE INTEGER EQUIVALENT OF : ' " + ORIGSTRING
                             + " ' IS : " + INTVAL ) ;
272                    }
273         }
274 }
```

Most of the code appearing in Listing 12.6 should be familiar to you—so I won't bore you by discussing the common elements. However, I do want to review the changes related to the revised design.

The test to determine which conversion to use begins with line 250:

```
IF ( CHARACTER . ISDIGIT (ORIGSTRING . CHARAT (0) )  )
```

This IF statement interrogates the first character of the string entered by the user.[39] If the *test condition* evaluates to TRUE, the first character is a digit, and the program assumes an Integer-to-Roman conversion and executes the code located on lines 255 through 265 (ignoring the comment lines). Otherwise, it performs a Roman-to-Integer conversion and executes the code appearing on lines 270 through 271. (Remember, the ELSE block of an IF statement executes when the *test condition* is FALSE.)

Compile and Execute

The screenshot depicted in Figure 12.7 displays the output resulting from the merged program's compilation and execution.

FIGURE 12.7   Merged program execution.

---

[39] Remember, we start counting at 0. Thus, the first character the user entered resides at location 0 in ORIGSTRING.

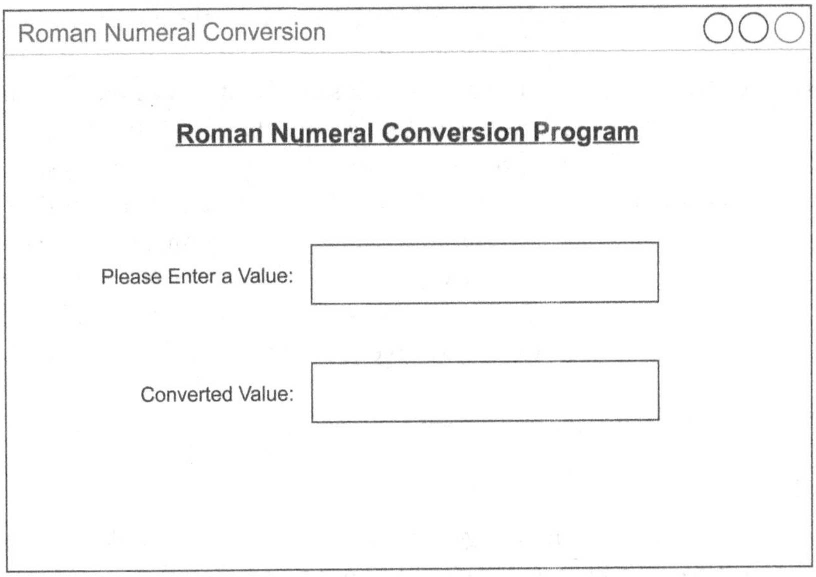

FIGURE 12.8 Simple GUI wireframe.

Additional Program Improvements

Although the merged version works reasonably well, I still wouldn't consider it a professional-caliber program (coding constructs aside). It would benefit from some additional improvements. I've listed several suggestions below.

GUI Interface Today, the user interfaces for most programs are graphical (GUIs). When designed well, GUIs are intuitive, straightforward, and simple to use. We could extend our program to accept and display values using graphics rather than line-oriented input. Figure 12.8 depicts a wireframe[40] representation of a simple GUI that would meet the need.

Looping Even if we were to retain the command-line interface, I think we could improve the user experience a bit. As currently designed, the program performs one conversion then exits, which might be annoying to users who need to convert more than one value.

To address this issue, we could modify the MAIN() method such that the prompt/conversion code resides within a loop. Thus, the user may convert as many values as they choose until entering a command that causes the program to exit. For example, we could add code to test

---

[40] *Wireframes* are minimalist mockups (i.e., sketches) of screen layouts. They help designers visualize the overall arrangement of interface components.

for the character "Q"[41] and have the program terminate when it receives it.

Extended Notation   Notational extensions exist that allow Roman Numerals to represent numbers greater than 3999. We could enhance our program to accept and process such values.

Input Validation   As currently designed, the program performs little validation of user-entered input. For example, this version of the program accepts malformed Roman Numerals.[42] Well-written, production-caliber applications validate all data to the extent possible to ensure accurate results.

If you're feeling ambitious, try implementing some of these enhancements.

SUMMARY

This chapter leveraged all the knowledge we gained thus far and developed two working programs. One converts a Roman Numeral to an integer; the other performs the complement operation. We also demonstrated how to combine both versions into a single executable. In addition, we presented several improvements to transform the code into a professional-caliber application.

Thus far, we have discussed the benefits of software and how computing technology improves and enhances our lives. Unfortunately, as with most aspects of life, there's a dark side. In Chapter 13, we'll review how nefarious individuals exploit this power for personal gain.

---

[41] This works nicely because "Q" is not a valid entry for either type of conversion. There are other options like having the user enter a blank line.

[42] Examples are left as an exercise for the reader.

Security and Privacy

It has become appallingly obvious that our technology has exceeded our humanity.

ALBERT EINSTEIN

I'd be willing to wager hard cash (or some Bitcoin) that most readers of this book rarely pay for software. For example, most of you likely get your email from one of the major providers, listen to music from one of the leading streaming services, and check your credit score with one of the primary reporting agencies. What's the common attribute? They're free. It can't get any better than that, right?

Well, have you ever wondered what the business model for these organizations might be? Consider that we've just spent the last twelve chapters of this book explaining how difficult and costly it is to design, develop, and deploy software systems. Do you think that the organizations sponsoring these "free apps" are spending all that money out of some altruistic largess? Or have they found some other way to "monetize" their investment?

Before we continue, I'd like to make a point: I'm not about to start bashing "big business" or the right to pursue profits. Free enterprise is the economic engine of liberty. Based on supply and demand, the market (companies and consumers collectively) determines the cost of goods and services, not some anonymous, unnamed bureaucratic governmental agency.

Overall, a free-market economy works well—but it's not without its challenges and abuses. As is always the case, *caveat emptor.*

Returning to the main point, how do organizations whose primary purpose is to pursue profits generate revenue when they practically give away their products and services? What is it that we're missing? The answer is: *you. Moi,* you say? Yes, *vous.*

These companies are selling *you*—or, more accurately, your *Personally-Identifying Information* (or, PII). How? They collect, store, and repackage your PII and market it to other organizations—which could be other corporations or even governments.

In the sections that follow, we'll present some of the obvious and devious ways organizations acquire and resell your PII—with or without your knowledge and consent. After reading this, if you become as angry as I am, all I can say is: Welcome to the club.

WHAT DO PEOPLE KNOW ABOUT YOU?

As noted above, your PII is a valuable commodity. But what exactly is it?

> *Personally-Identifiable Information* (PII) is data that either alone or in conjunction with other data elements can uniquely identify a specific individual.

The following list is but a sample of the data elements collected about you. I don't want to sound like an alarmist, but it's likely that much of this information is already available for purchase from legitimate vendors or is otherwise accessible from more nefarious suppliers operating in the dark shadows of the Internet. (More on this later.)

Please note that this list is by no means exhaustive.

Name	Current full name, former name (e.g., maiden), nicknames, your mother's maiden name
Demographic Data	Date of birth, race, religion, marital status
Contact Information	Telephone numbers, including mobile, home, and business
Personal Attributes	Education Levels, Degrees, security clearances, pet ownership
Location Data	Home street address, email address, PO Box numbers, city of birth
Relationships	Family members, friends, business associates
Associations	Employers, employment history, clubs, organizations, volunteer groups
ID Numbers	Social security number (scary!), driver's license number (also scary!), license plate tag, passport number, credit card numbers, financial account numbers, electronic toll collection account IDs
Digital Data	The IP and MAC addresses of the computing devices you commonly use (e.g., your phones and laptops)
Physical Characteristics	Photographs (digital images), biometric data (fingerprints, retinal scans, etc.)
Financial Assets	Banking data, property ownership, owned or leased vehicles (VINs), credit cards, loans, debts, bankruptcies

Speaking for myself, I find the scope and extent of this list disturbing at best, terrifying at worst.

WHY SHOULD YOU CARE?

Why should you care? At a high level, that seems like a straightforward question with a straightforward answer: because you're the one that pays the price when something goes awry. The burden falls on you to reconcile accounts, repair your credit, get refunds for products you didn't purchase, etc.

But the actual answer to that question is a bit more subtle. Why? Because your PII data is often used without your knowledge and consent and in ways that are not so obvious.

The sections that follow describe how this happens.

General Concerns

Does it bother you that:

- When you walk into many casinos, the staff knows who you are before you place your first bet?

- Search engines know more about your life than your family does?

- When you drive, traffic cameras and electronic toll devices trace your trip?

- Your phone continually tracks your location?

- Social media outlets can accurately determine sexual orientation, race, relationship status, political leanings, etc.?

Have you experienced any of the events described below?

- When you and your spouse discuss some shopping needs within "earshot" of your smart device (e.g., Alexa or Google Home), did you suddenly receive email ads for those products?

- When you window shop in front of a store in a mall, do discount coupons arrive as alerts on your phone?

Have you considered these issues?

- How can you determine whether every cell tower your mobile phone connects to is part of your provider's network? Could a government (the USA or otherwise) or a third-party data collector own and manage some random cell towers?

- Wi-Fi and cellular networks are becoming so accurate in determining your phone's location that they'll be able to identify which products you're browsing in a store.

- Do you allow your cell phone to connect to public Wi-Fi networks? If so, do you know whether unknown third parties are monitoring your data traffic?

- Are you concerned that identity thieves can file false tax returns in your name? These scammers enter false deductions to ensure a refund and then have the IRS wire the payment into one of their bogus bank accounts. Try getting that money back.

I could continue, but I'm sure you get the point. You may think I'm paranoid, but I assure you that every one of the above examples is real. And they happen every day. Sadly, I'm sure some readers of this book have been victims of one or more of these attacks—including me.

In addition to the general concerns noted above, there are specific issues outlined in the next few sections.

Monetization of Data

The title of this section sounds very professional and businesslike. But what it really means is that companies you do business with every day—cellular providers, electricity suppliers, and banks, to name a few—sell your data often without your knowledge and consent. But what galls me is that it's *my* data, and I don't get to authorize who gets it, nor do I receive any remuneration for it.

Data Breaches

Organizations lust after your PII—as much of it as they can get their digital digits around. And sometimes you have no choice but to give it to them. For example, it's impossible to apply for a credit card, register a car, or receive medical care without providing a wealth of personal information.

Unfortunately, once you provide your PII to some third party, you're no longer in control of it—it's in someone else's hands. You must rely on them to protect your data as vigorously as you would.

But do they?

Regardless of their diligence and competence, security professionals must design systems to protect your data 24×7×365. A hacker needs to get lucky only once.

Right now, unbeknownst to you, your data might be for sale to anyone willing to pay for it. Moreover, there's no way to anticipate how criminals might use it—they're *highly creative*. If you think I'm exaggerating, just launch your favorite search engine[1] and query "data breaches." The length of the list returned is staggering. Sadly, whether you're aware of it or not, it's likely you've been affected by one or more of them.

Identity Theft

Identity theft is rampant, but many folks believe it's limited to credit card fraud. Unfortunately, it's much more pervasive and pernicious than that. Given enough of your PII—much of which might already be available on the Web—cybercriminals can create a "shadow identity" that looks and acts like you to the rest of the world. They can apply for credit cards, passports, driver's licenses—anything they want in your name. When the dust settles and the payments come due, who do you think will be left holding the bag?

Malware

The term *malware* refers to any code written for malevolent purposes. It disguises itself in many forms. Some common examples appear below.

Adware Adware is an aggressive form of advertising that typically presents itself as annoying popups; they are more irritating than malicious. However, if you see them, it's likely that your computer has been compromised and that other, more

---

[1] We'll discuss the shortcomings of search engines later in this chapter.

malevolent forms of malware may piggyback their way into your system.

Viruses   A computer virus inserts itself into "clean" files and then attempts to infect other files. They are so named because their behavior imitates their biological counterparts. Unfortunately, they do more than just replicate. They also delete files, corrupt folders, and damage systems.

Spyware   As its name might suggest, the intent of spyware is to snoop—on you. It lurks in the background and surreptitiously collects data such as logins, passwords, and account numbers. It then packages and sends this data to its "creator," who will gladly make it available for purchase.

Ransomware   This might be the most heinous form of malware. Ransomware assumes full control of your computer. It then issues an ultimatum: pay a ransom, or it will delete all your files and render your system unusable.

HOW DO ORGANIZATIONS ACQUIRE DATA?

If nothing else, humans are incredibly innovative. You see evidence of this in museums and theaters, hear it in music and poetry, and smell it in the kitchens of Michelin four-star restaurants. Sadly, however, ingenuity is not the sole province of kindly, well-intentioned people. As we will see, some of the most creative minds you've never heard of focus their faculties on finding inventive ways to acquire and use your PII.

Free Services

It may surprise you to learn that organizations have developed a very inventive way to acquire data from you: they ask you for it. I don't want to belabor this point, but companies must generate revenues from somewhere to remain viable and underwrite the costs associated with employees, servers, offices, electricity, etc.

As we discussed above, there are no free lunches. So be very wary of any complimentary service you see marketed on the Internet. The provider will ask you many innocent and helpful-sounding questions when you sign up. But I assure you, it's not because they care about you; on the contrary, they need something to sell to maintain their revenue stream.

There are many means—both obvious and devious—by which organizations track you online. Let's unveil some of them.

Tracking Your Surfing

Most users would find it annoying having to enter a password every time they press a button on a web page. Thus, websites save session data in files called *cookies*, and your browser automatically includes this information with every message it sends to the webserver.

In addition to retaining authentication data, developers use *cookies* to maintain lists of items recently viewed, products placed in shopping carts, and previously entered data (such as a credit card number). Very convenient, to be sure.

Unfortunately, everything that can be exploited will be exploited.[2] *Cookies* are no exception to that rule because they can be used to create *tracking IDs* that uniquely identify you (or your device). Using this ID, advertisers (and other nefarious organizations) can monitor browsing history and target advertising.

How targeted advertising works is both simple and ingenious. When you visit a website, say SOME-WEBSITE.COM, you might see an ad positioned somewhere on the web page. That random, innocuous-looking advertisement is associated with its own website, let's say, SILENT-TRACKER.COM which, in its *cookie*, will record the fact that you visited SOME-WEBSITE.COM. (To be clear, the website associate with the ad—SILENT-TRACKER.COM—records in its *cookie* that you visited SOME-WEBSITE.COM.). Later, when you visit other websites, that same ad may appear, and the same action takes place: SILENT-TRACKER.COM saves each site you visit in its *cookie*.

As this process continues and data accumulates, advertisers can paint a surprisingly good picture of your life and lifestyle: you're now subject to targeted advertising. For example, let's say you're shopping for a new car, and you visit several dealership websites. You may start getting ads for other vehicle models, automobile insurance, and bank loans.

Device Fingerprinting

Because techniques to impede tracking *cookies* are becoming prevalent (see below), advertisers need to develop more subtle methods to monitor you. One approach gaining popularity combines such data elements as your IP address, location information, system configuration—and sophisticated algorithms—to create what's called a *device fingerprint* that uniquely identifies your system. Thus, advertisers can track all activity originating from your computer. (Did you ever agree to this? I know I didn't.)

Device fingerprints are just as effective as tracking IDs but are far more difficult to thwart because the data used to create them is unavoidably visible.

Social Media

Social media giants—Facebook, Twitter, WhatsApp, YouTube, Instagram, etc.—likely know more about you than your lover. Secrets you wouldn't even hint at to close friends or family are easily inferred based solely on your "click" history: postings, likes/dislikes, video selections, etc. Furthermore, they aggregate this data with information they've gathered from other sources, such as credit reports, web searches, and shopping tendencies, to build a complete and all-too-accurate picture of you.

Armed with this knowledge, these organizations can accurately predict personal attributes such as sexual orientation, political leanings, legal and illicit drug use—the list is endless. In addition, they can accurately predict that you're going to buy a particular make and model of a car before you even walk into the showroom, know that you're expecting a child before you announce the good news to the world, and correctly forecast how you'll vote in the next election. What's worse—at least in my opinion—is that they sell these predictions without your knowledge or consent.

---

[2] A wise person once said that. Well, to be honest, it was me and I'm not so wise—I'm simply tired of feeling exploited.

I find such tracking and predicting both alarming and creepy. So much so that I don't have any social media accounts. The benefits these sites provide don't outweigh what I believe is an outrageous and shameful violation of my privacy.

HOW CAN YOU PROTECT YOURSELF?

To protect yourself, start by accepting that you're a target and that you may have already fallen victim to many of the attacks discussed above. Once you've adopted that mindset, you can use the techniques described below to make it difficult for organizations to invade your privacy.

Think Before You Share

The title of this section speaks volumes: don't give away personal data thoughtlessly. Just because a website requests information doesn't mean you *must* provide it. Before you share your PII, ask yourself whether the organization seeking it truly needs it to provide the service in question. If not, say "no."

If it's a "free" site and the provider won't take "no" for an answer (i.e., they won't grant access without acquiring your PII), pull your hands away from the keyboard and ask yourself, is it worth it? Do I really need this service? If the answer is "yes," you still have two options: search for another provider or find a fee-based alternative that purports to protect your privacy. (Consider spending a little money now to avoid significant issues in the future.)

If you opt to use a free service, keep in mind that, even if the provider appears socially responsible, they still glean knowledge about you based on your interactions with their site. Every one of your "clicks" says something about you. As painful as it is, read the Terms of Service (TOS). That's the only way to understand the level of risk you're assuming.

Once you've completed the registration process, immediately locate the privacy settings and opt-out of every data collection option you can find. For example, many sites ask if they can track and share usage data; say no. Other sites ask if they can monitor activity to improve their product or service: say no. If a product—typically browsers—provides an option to delete cookies after a session terminates, say yes. Scrutinizing such detail is distressingly tedious—but I assure you that identity theft is far more agonizing and painful.

As an aside, this advice holds true in the legacy world of paper as well. When you enter data on forms, questionnaires, and documents, view every field as a *request* for your information. Feel free to say no.

Whenever and however you provide personal data, you're relying on a third party to protect it. If they experience a security breach, it's a bad day for them—but it could be a bad year for you.

Protect Passwords

We live in an extraordinary age of digital convenience. We shop, bank, and communicate with the press of a button—or by verbal command. But there are risks.

Ask yourself: Would you ever (deliberately) leave your house keys where they could be stolen or copied? Of course not. But most folks don't recognize the fact that passwords are just as important.

Hackers and cybercriminals are continually trying to gain access to your accounts. They are smart, motivated, and use powerful computers customized for this purpose. Moreover, they have all the time in the world and only need to get lucky once.

Don't be a victim. Below are some guidelines for protecting passwords. Please note that by no means should you consider this an exhaustive list.

- Passwords should be at least twelve characters long, contain upper and lowercase letters, and include numbers and symbols (e.g., "$" and "+").

- Don't jot down passwords on post-it notes and stick them to your monitor.

- Don't choose passwords that relate to you in any way. For example, don't use your dog's name, your birth date, or the title of your favorite song. (As per our earlier discussion, cybercriminals can acquire such information from social media sites and use it to compromise your passwords.)

- Never reuse a password. Ever.

- Never use passwords for more than one site. Remember that, through tracking IDs, hackers may know all the websites you visit. A password for a closed account should not allow access to an active one.

- Don't use formulas to create passwords. For example, if you have an account at SOMESITE.COM, don't create passwords of the form: SOMESITE123 or ABCSOMESITE. Though easy to remember, passwords generated in this manner are a hacker's dream.

- Change passwords frequently and routinely. Yes, this is painful, but it's worth it.

- Unless there's a compelling reason, never disclose passwords to anyone—including work or school colleagues. On those rare occasions when you must reveal a password, change it as soon as you can.

- If your device supports biometrics, use them. Fingerprints are unique, require no memorization, and travel with you.

Log Off

Take the extra moment to logout. When you initiate a session with a website, servers often store (or *cache*) some of your data in memory—rather than on a secure disk—so that it can respond to subsequent requests quickly and efficiently. This data is often stored *en clair*. That is, it's not encrypted and thus subject to hacking. When you logout, the server clears its cache and removes your data from memory.

Browsers and Search Engines

There are several popular browsers and search engines available today. Like any product, each offers different levels of service and privacy. I suggest you use a browser and search engine combination that vigorously safeguards your privacy and doesn't facilitate the use of tracking IDs (see the discussion above).

Once you select a browser, there are several options you should configure. Each product is unique, but here are some general tips to protect your privacy.

- Clear the *cache* and *cookies* on exit.

- Prevent the creation of *tracking IDs* and device *fingerprinting.*

- Inhibit social media tracking.

- Install third-party add-ons from trusted providers that block ad trackers.

- Use *private browsing.* This option prevents recording your browsing history on the device you're using. However, this is not a panacea; the rest of the Internet can still follow your every move. For greater privacy, use a VPN (see below).

Wi-Fi Networks

Do not use public Wi-Fi networks. Ever. Anyone could be eavesdropping on the connection. Pay the data charges for a private service. Trust me: it's cheaper in the long run.

Update Software

For most software products, every update contains security improvements. It's a constant cat-and-mouse game: hackers identify new flaws; vendors repair them. Your best defense is to stay current.

Anti-Malware Software

Purchase and install *anti-malware* software. The objective of these products is to prevent, detect, and delete all types of malware. Vendors continually update their products whenever new risks emerge. I fervently encourage you to invest in one. I know it's easy for me to spend your money but consider what the cost might be if you don't.

Use a VPN

As discussed in Chapter 7, computers must connect to an Internet Service Provider (ISP) to access the Web or communicate with devices outside your home or office network. That means that your ISP can scrutinize, filter, or even censor your traffic.[3] It also means that third parties—such as advertisers and governments—can also monitor your activity.

One way to increase privacy is by using a personal Virtual Private Network (or VPN).

As depicted in Figure 13.1, VPNs create an encrypted connection—called a *tunnel*— between your device and a VPN server. To do that, VPN software installed on your computer encrypts all message traffic before directing it to a VPN server.

Upon receipt, the VPN server decrypts the message and forwards it *en clair* to the intended destination. However, the IP address visible on the Internet is that of the VPN's

---

[3] There seems to be *Big Brothers* everywhere you look these days.

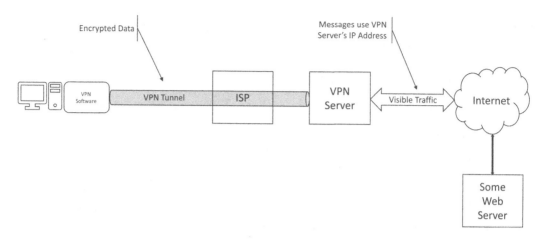

FIGURE 13.1   VPN topology.

server, not yours. In this manner, *VPN Tunneling* masks your identity and location, and makes it difficult to track your activity.[4]

Verify Invoices and Bills

This suggestion may seem obvious, but carefully peruse invoices, receipts, credit card bills, and bank records. Unexpected entries on statements may be your first indication that cybercriminals have compromised an account or stolen your identity.

If there's any doubt whatsoever, apply some tourniquets and stop the bleeding as quickly as possible: change passwords, lock accounts, and call financial institutions immediately.

Clean Up

Delete (close) unused accounts. Though dormant, they still contain all your PII and remain targets for a data breach. Moreover, it might take a while to realize that hackers are wreaking havoc with your information because you pay no heed to inactive accounts.

Remain Wary

Hackers and cybercriminals are after your data. That's a fact—not paranoia. Unfortunately, despite the constant threat, we receive precious little help from governments and police departments.

In the end, it's up to you to protect yourself. Below are some guidelines I suggest you follow:

- If you see an email that looks *hinky*, delete it without opening it. If you think it might be from someone known to you, ask the sender if the message is legitimate.

- Never "click" on a link contained in an email from an unknown sender. Doing so is just asking for trouble.

---

[4] Nota Bene: If your VPN service provider maintains activity logs, your browsing history might be available *post facto*.

- In a similar vein, don't click any buttons on any websites that are not secure.

- Never open an email attachment sent by somebody that you don't know. If you're unsure, delete the message, and ask the sender to resend the attachment.

- Type important URLs directly into your browser—don't rely on links sent to you. Cybercriminals can mask URLs. Thus, you might think you're interacting with your bank's website when, in reality, you just typed your username and password into a hacker's database.

- Don't download anything from a questionable website. Ever.

- Install anti-malware and anti-virus software and never turn it off.

ADVANCED TOPIC: THE DARK SIDE

As discussed earlier in the book, the Internet is nothing more than an interconnected set of independent networks that adhere to the TCP/IP protocols. That said, anyone can connect any computer to any Internet-based network at any time. However, not every connected device is part of the World Wide Web.

For example, consider that laptops and smartphones gain access to the Internet by connecting to an ISP's network. However, does that mean every Web-based system can locate these devices? It's not likely because your ISP assigns them *temporary* IP addresses[5] that are valid only for the current session and are not known beyond the boundaries of your local network.

In general, for a server to be visible on the Web, Search Engines, such as Google, Bing, DuckDuckGo, etc., must know of its existence.[6] That begs the question: What's a Search Engine?

> A *Search Engine* is a software application comprising two main parts: a *cataloging component* that scans the Web,[7] indexing every data element found on every website, and a *searching component* that, based on user queries, displays the URLs of previously indexed websites.

Search engines organize and catalog vast amounts of information and provide powerful query tools that allow users to locate data, sites, and services with minimal effort. It's not hyperbole to state that they've become the *sine qua non* of the Internet. Indeed, consider that they've also changed the cultural landscape: we no longer "search for stuff"—we *Google* it.

However, remaining "off the grid" is often beneficial. For example, consider vendors that offer paid subscriptions for news, games, movies, proprietary research, etc. Such sites will often deploy public-facing webpages indexed by search engines but keep their content servers private. Users would only acquire the URLs for the content servers after rendering payment.[8]

---

[5] ISP-assigned IP addresses may change with every session.

[6] Getting your webpage indexed by search engines "puts it on the map." This is important for any site that wants traffic directed to it. In addition, there are other techniques by which you can share IP addresses.

[7] IT professionals refer to this process as *crawling the Web.*

[8] We say that such servers reside behind *paywalls.*

This collection of unindexed servers forms what's called the *Deep Web*.

The *Deep Web* comprises all unindexed servers connected to the Internet.

Most of the Deep Web is benign and supports constructive uses such as online access to financial accounts, medical data, and business-to-business transactions. But, as I stated previously, everything that can be exploited will be exploited.

To that end, there are websites on the Deep Web that host more nefarious activities such as drug trafficking, illegal weapon sales, marketplaces for stolen PII—and much worse. This subset of servers forms the *Dark Web* or *Darknet*.

The *Dark Web* (or *Darknet*) is a subset of the Deep Web that requires specialized software, custom configurations, and explicit authorization to access any of its servers.

Using specialized software,[9] users can remain anonymous on the Dark Web. As a result, cybercriminals can conduct their nefarious transactions in the shadows of cyberspace. When hackers acquire your PII, this is where they go to sell it to the individuals who will exploit it.

For the casual user, the Dark Web is not a place to visit. Despite any curiosity you might have, I'd urge you to steer clear of it.

SUMMARY

In this chapter, we discussed some of the darker aspects of the Internet. We learned that cybercriminals target anyone and everyone and that individuals must undertake adequate precautions while surfing the net. To that end, the chapter outlines many ways users can protect themselves.

We also described the Wild West of the Internet: the Deep Web and its more nefarious subset, the Dark Web. It's the Dark Web that facilitates some of the most heinous activities that occur on the Internet.

---

[9] This is *well* beyond the scope of this text.

Glossary of Terms

Abstraction Abstraction is selecting the essential and relevant attributes of a problem domain to simplify its representation.

ADC See Analog-to-Digital Converter.

AI Artificial Intelligence.

Algorithm An algorithm is a problem-solving technique suitable for implementation as a computer program.

ALU—Arithmetic/Logic Unit The ALU (Arithmetic/Logic Unit) is a subcomponent of the CPU and is responsible for executing most of the system's instructions.

Analog-to-Digital Converter Analog-to-Digital Converters capture and transform sound waves into discrete digital values.

Application An application is a finite arrangement of instructions packaged as one or more programs that, when executed, use data to accomplish a specific, targeted set of tasks within a computing environment (see Program).

AR Augment Reality.

Architecture See Software Architecture.

Authentication Determine/Validate a user's identity.

Authorization The list of actions the system permits a given user to perform.

Backbone A backbone is a network connection that joins multiple network segments.

Biometric Devices Biometric Devices (such as fingerprint readers, retinal scanners, and facial scanners) provide convenient and secure user authentication mechanisms based on physical characteristics.

BIOS Basic Input/Output System; see Firmware.

Boolean Expressions Boolean expressions (or "Booleans") are logic-based operations that yield values of either "TRUE" or "FALSE."

Bridge A bridge is a component that interconnects two networks.

Bus The bus transports data among a computer's internal components.

Bytecode Bytecode is a machine-like instruction set optimized for execution by a software interpreter or virtual machine. (See JVM.)

CAN Campus (Corporate) Area Network.

Casting Casting is a way to project an image appearing on a small display (e.g., a smartphone's screen) onto a device with a larger format such as a TV or computer monitor.

Client-Server Model The client-server model is a communications framework that pairs service requestors with service providers via a network connection.

Code Blocks Programming blocks allow developers to group multiple statements and treat them as a single unit. (See also: Named Code Blocks.)

Command-Line Interface A user interface consisting primarily of text-based input and output. (See also: GUIs.)

Compilation Compilation is the process of translating a computer program written in one language (called the source language) into another language (called the target language).

Compiler A computer program that performs compilation. (See compilation.)

Computer Programming Computer programming is a set of tasks and procedures conducted by software developers that, when completed, generate a set of executable instructions that achieve a specific result.

Computer Virus See Virus.

Context Switching Context switching is the saving and restoring of a process's execution context.

Control Unit See CU

CPU The Central Processing Unit is the "brain" of a computer.

CU Control Unit—A subcomponent of the CPU.

DAC See Digital-to-Analog Converter.

Dark Web The Dark Web is a subset of the Deep Web that requires specialized software, custom configurations, and explicit authorization to access any of its servers.

Data Flow Diagram A Data Flow Diagram (or DFD) captures how data travels through a system.

DAW Digital Audio Workstation.

Deep Web The Deep Web comprises all the unindexed servers connected to the Internet.

Design See Software Design.

Digital-to-Analog Converter A Digital-to-Analog Converter transforms digitized audio data into waves suitable for playing through speakers or headphones.

Digitization Digitization is the process of converting data into a binary representation that consists solely of 1s and 0s.

DNS Domain Name Service—Maps Domaine Names to IP Addresses.

Encapsulation Encapsulation is a design methodology wherein all implementation details (data and code) of a component remain private. Programmers only know—and interact with—the component's public interface (i.e., its methods and functions).

EPN Enterprise Private Networks.

Ethernet Ethernet is a network protocol that packages, manages, and controls data transmission (bits) over LAN connections.

Execution Context Execution Context is the state of all system resources for a given process at a given moment in time.

Extranet An Extranet is an Intranet that permits controlled (and often limited) access to external entities.

Firewall A firewall is a component that prevents unauthorized access to a network.

Firmware Firmware is software integrated into the hardware that controls the operation of the underlying circuitry.

Flash Drives Flash Drives (sometimes called thumb drives) are portable disk drives that allow users to store and transfer files from one machine to another.

Flash Memory Flash Memory is a type of chip that can retain its state (data) in the absence of power.

FPU Floating-Point Unit; see Math Coprocessor.

Functional Requirements Functional Requirements specify the features the application must provide.

Fully Connected Mesh Network In a fully connected mesh network, every node connects to every other node.

Gateway A gateway is a networking device that interconnects otherwise discrete networks to each other.

GUI Graphical User Interface is a user interface that presents icons and graphic to the user. (See also: Command-line interface.)

HAN Home Area Network. (See LAN.)

Hard Drive A disk drive that stores data magnetically on specially coated rotating platters.

HTTP Hypertext Transfer Protocol.

Hub A hub is a communication device used to construct a private network.

Hybrid Network A hybrid network comprises multiple topologies.

IDE IDEs, or Integrated Development Environments, incorporate editors, compilers, and debuggers into a single development dashboard.

Interpreter An interpreter is a computer program that directly executes instructions without requiring compilation.

Interprocess Communication Interprocess Communication (or IPC) is the ability for two (or more) running processes to exchange data.

Interrupts Interrupts are signals from hardware devices (e.g., disk and network controllers) indicating that they need attention from the OS.

Intranet One or more networks controlled one organization.

IoT The Internet of Things.

IP Internet Protocol.

IPC See Interprocess Communication.

ISP Internet Service Provider.

JVM Java Virtual Machine—this is a utility that interprets and executes Java bytecode as generated by a Java compiler.

Kernel Mode When running in kernel mode, the OS has unrestricted access to all system resources. (See also User Mode.)

LAN Local Area Network.

MAC Address A unique identifier assigned to every network interface card (NIC).

Main Memory Main Memory is the primary internal storage within a computer. Every program's instructions and data must reside in memory before the CPU can begin its execution.

Malware Software deliberately designed to cause damage.

MAN Metropolitan Area Network.

Math Coprocessor A specialized component to perform operations on floating-point numbers.

Mesh Network See Partially Connected Mesh Network and Fully Connected Mesh Network.

Method A method is a named group of statements that collectively perform a specific function.

Microcode Microcode is a layer of software that executes the machine instructions in the hardware.

MIDI Musical Instrument Digital Interface.

Model-View-Controller Architecture Model-View-Controller is a software architectural model comprising three significant subcomponents: The Model (the data subcomponent), the View (the user interface subcomponent), and the Controller (the subcomponent that updates the View and maintains the integrity of the Model).

Modem A modem, or MOdulator-DEModulator, allows network traffic to flow over transmission media not originally intended to support digital traffic (e.g., telephone lines).

Modular Design Modular Design is a design technique wherein developers decompose functionality into cohesive, independent components, called modules. Each module contains all the code required to perform one feature of the system.

Module A self-contained unit of functionality.

MTA Mail Transfer Agent.

MUA Mail User Agent (an email application).

MVC See Model-View-Controller Architecture.

Named Code Blocks Named Code Blocks allow programmers to repeatedly invoke blocks of code by name rather than inserting copies each time. Examples include subroutines, procedures, functions, and methods. (See also: Code Blocks.)

Network Interface Controller A device used to connect a computer to a network.

Networking Networking is the ability to connect two (or more) electronic devices with the express intent to share data and computing resources.

Network Topology Topology characterizes the organization and hierarchy of the nodes residing on a network.

NIC See Network Interface Controller.

Node A component on a network.

Non-functional Requirements Non-functional Requirements specify such items as performance, availability, and reliability, etc.

OCR See Optical Character Recognition.

Opcode An opcode (or Operation Code) is a machine-language instruction that specifies the operation (e.g., add, subtract, copy, move) that the CPU will execute.

Operating System An Operating System is a specialized program that oversees process execution and manages all the hardware and software resources available in a computer system.

Optical Character Recognition Optical Character Recognition (OCR) software scans images and converts text (both printed and handwritten) into its equivalent machine encoding.

Optical Drive External storage devices that use lasers to read and write data.

OS See Operating System.

OSI The Open Systems Interconnection Network Model.

OS Trap An OS trap is a synchronous event that requires the operating system to intervene.

PAN Personal Area Network.

Parameters Parameters serve place holders for values passed to named code blocks (e.g., subroutines, functions, procedures, and methods).

Partially Connected Mesh Network In a partially connected mesh network, every node is aware of every other node on the network and tries to establish a direct connection with as many of them as possible.

Peripheral Devices Peripheral Devices are external components that connect to computers.

Personally Identifiable Information Personally Identifiable Information (PII) is data that either alone or in conjunction with other data can uniquely identify a specific individual.

PII See Personally Identifiable Information.

PITA See Point-in-Time Application.

Point-in-Time Application A Point-in-Time Application integrates multiple components to satisfy a specific user request at a given moment.

Process A process is the running image of a program as it executes in memory. (See program.)

Program A program is an ordered collection of instructions contained in a file. (See process.)

Programming Language A programming language is a formalized set of general-purpose instructions that produce a precise result when combined and executed in a specific arrangement.

Pseudocode Pseudocode allows developers to describe programming logic using human-readable constructs without becoming mired in a specific language's syntactic details.

QR Codes QR codes (or Quick Reference codes) are a variant of barcodes.

RAM Random Access Memory.

Ransomware Ransomware is an extreme form of Malware that takes over an entire system and then demands a "ransom" to unlock it.

Register A high-speed memory location residing in, and controlled by, the CPU.

Requirements Specification A Requirements Specification defines and delineates the precise set of functions and features that a given software system must contain.

RFC RFC Stands for "Request for Comment." When IT professionals want to share a new idea, they write an RFC and distribute it. Other IT professionals read it and then comment and suggest changes. Eventually, if enough experts converge on an agreement, the RCF becomes the basis of a new specification.

Ring Topology In a ring topology, every node connects to two nodes in sequence, thus forming a ring.

ROM Read-Only Memory.

Router The primary purpose of a network router is to forward data packets between two or more networks.

SAN Storage Area Network.

Scheduling Scheduling is the procedure by which the OS allocates system resources to running processes.

SDLC See Software Development Lifecycle.

Search Engine A Search Engine is a software application comprising two main parts: A cataloging component that "crawls" the Web, indexing every data element found on every website, and a searching component that, based on queries entered by users, displays the URLs of previously indexed websites.

Service A Service is a unit of independent, self-contained functionality that performs one task, and its implementation is opaque (i.e., anyone using it doesn't have to know how it works).

Service-oriented Architecture A Service-oriented Architecture is a design approach wherein individual services are available via network connections to all authorized application components.

Software Software is an organized arrangement of data and instructions that, when executed, direct the operations of the underlying hardware to accomplish a particular purpose.

Software Architecture Software Architecture defines the major components (also called modules) that comprise an application. It also establishes how these components interact.

Software Design Software Design is the process of creating specifications for modules that adhere to architectural requirements.

Software Development Lifecycle The Software Development Lifecycle is the management framework required to plan, design, develop, and deploy computer systems.

Solid-State Drives Solid-state drives use integrated circuits to retain state (data) in the absence of power.

Speech Recognition Speech recognition is the process that converts spoken language into digital text suitable for processing by electronic devices.

Spyware Spyware surreptitiously inserts itself into systems to collect data such as logins, passwords, and account numbers.

SSD See Solid-State Drives.

String In Java, a STRING is a grouped sequence of characters.

Switch A communication device used to form a private network.

TCP Transport Control Protocol

TCP/IP TCP/IP is one of the most common networking models in use today.

Thumb Drive See Flash Drive.

Topology See Network Topology.

Trap See OS Trap.

URL Universal Resource Locator.

USB Universal Serial Bus.

User Mode When running in user mode, the system prohibits application processes from acquiring access to any protected system resources. (See also: Kernel Mode.)

Virtual Machine A Virtual Machine is a software representation of a computer system that emulates all the underlying hardware functionality.

VR Virtual Reality.

Virus A computer virus is malicious software that inserts itself into a "clean" file and attempts to infect other files. They are so named because their behavior imitates their biological counterparts.

VPN Virtual Private Network. (See VPN Tunneling.)

VPN Tunneling A VPN Tunnel is a secure, encrypted connection between a source system and a VPN server.

WAN Wide Area Network

Reading List

HISTORY OF COMPUTING

- Ceruzzi, P. E. (2003). *A History of Modern Computing*, Second Edition. The MIT Press.
- Williams, M. R. (1997). *A History of Computing Technology*, Second Edition. Wiley-IEEE Computer Society.
- Ifrah, G. (2001). *The Universal History of Computing: From the Abacus to the Quantum Computer*. Wiley.

COMPUTER HARDWARE

- Hsu, J. Y. (2001). *Computer Architecture Software Aspects, Coding, and Hardware.* CRC Press.
- Patterson, D. A.; Hennessy, J. L. (2013). *Computer Organization and Design (MIPS Edition)*, Sixth Edition. Morgan Kaufmann.
- Dumas II, J. D. (2017). *Computer Architecture Fundamentals and Principles of Computer Design*, Second Edition. CRC Press.

COMPUTER NETWORKING

- Hura, G. S.; Singhal, M. (2001). *Data and Computer Communications: Networking and Internetworking.* CRC Press.
- Morreale, P. A.; Anderson, J. M. (2015). *Software Defined Networking Design and Deployment.* CRC Press.
- Ross, K. W.; Kurose, J. F. (2017). *Computer Networking: A Top-Down Approach*, Sixth Edition. Pearson India.

OPERATING SYSTEMS

- Comer, D. (2020). *Operating System Design: The Xinu Approach*, Second Edition. Chapman and Hall.
- Tanenbaum, A. S. (2016). *Modern Operating Systems*, Fourth Edition. Pearson India.

PROGRAMMING LANGUAGES

- Gabbrielli, M.; Martini, S. (2010). *Programming Languages: Principles and Paradigms.* Springer.

- Bansal, A. K. (2013). *Introduction to Programming Languages.* Chapman and Hall/CRC.

- Louden. K. C.; Lambert, K. A. (2011). *Programming Languages: Principles and Practices (Advanced Topics)*, Third Edition. Cengage Learning.

COMPUTER PROGRAMMING

- Chemuturi, M. (2018). *Computer Programming for Beginners: A Step-By-Step Guide.* Chapman and Hall/CRC.

- Thomas, D.; Hunt, A. (2019). *The Pragmatic Programmer: Your Journey To Mastery*, Second Edition. Addison-Wesley Professional.

- Althoff, C. (2017). *The Self-Taught Programmer: The Definitive Guide to Programming Professionally.* Self-Taught Media.

ALGORITHMS

- Bowman, C. F. (1994). *Algorithms and Data Structures: An Approach in C.* Oxford University Press.

- Louridas, P. (2020). *Algorithms.* The MIT Press.

SOFTWARE APPLICATION DESIGN

- Petre, M., Van Der Hoek, A. (Eds.) (2019). *Software Designers in Action: A Human-Centric Look at Design Work.* Chapman and Hall/CRC.

- Chemuturi, M. (2018). *Software Design: A Comprehensive Guide to Software Development Projects.* Chapman and Hall/CRC.

- Budgen, D. (2021). *Creating Solutions for Ill-Structured Problems.* Chapman and Hall/CRC.

SOFTWARE DEVELOPMENT METHODOLOGIES

- Beck, K.; Andres, C. (2004). *Extreme Programming Explained: Embrace Change*, Second Edition. Addison-Wesley Professional.

- Stellman, A.; Greene, J. (2013). *Learning Agile: Understanding Scrum, XP, Lean, and Kanban.* O'Reilly Media.

- Kung, D. (2013). *Object-Oriented Software Engineering: An Agile Unified Methodology.* McGraw-Hill Education.

COMPUTER SECURITY, PRIVACY, & MORALITY

- Kaczmarczyk, L. C. (2011). *Computers and Society Computing for Good.* CRC Press.

- Gould, C. C. (2019). *The Information Web: Ethical and Social Implications of Computer Networking.* Routledge.

- Manjikian, M. (2017). *Cybersecurity Ethics: An Introduction.* Routledge.

ARTIFICIAL INTELLIGENCE

- Russell, S.; Norvig, P. (2020). *Artificial Intelligence: A Modern Approach,* Fourth Edition. Pearson.

QUANTUM COMPUTING

- Johnston, E. R.; Harrigan, N.; Gimeno-Segovia, M. (2019). *Programming Quantum Computers: Essential Algorithms and Code Samples.* O'Reilly Media.

Index

Printed in the United States
by Baker & Taylor Publisher Services